Contents

1. Introduction — 1
   1.1. Methodology:
        Debating Immigration in an Imperfect World — 7
   1.2. Terminology:
        Refugees, Asylum Seekers, and Other Migrants — 12
   1.3. Setting the Empirical Stage:
        The Economic and Social Effects of Migration — 16

Part I: Normative Foundations: Basic Rights and the Ethics of Immigration

2. Basic Human Rights and Correlative Duties — 27
   2.1. The General Duty (I) to Avoid Violating Human Rights — 33
   2.2. The Assigned Duty (II) of States to Protect Human Rights and the Right to Self-Determination — 34
   2.3. The Duty (III) of States to Assist People Deprived of Basic Rights — 43
   2.4. Discussing Rights and Duties from a Risk-Ethical Perspective — 49

3. Toward a Right to Exclude Migrants — 52
   3.1. Rejecting the Cosmopolitan Case for Global Free Movement — 53
   3.2. The Importance of Community and the Need for Exclusion — 69
   3.3. Special Duties and the Right to Protect: On the Function of Social Trust in a System of Moral Division of Labour (MDL) — 80
   3.4. Special Duties and the Right to Control: Ryan Pevnick's Associative Ownership View — 84
   3.5. On the Special Claims of Migrants — 88

4.	The Special Case of Refugees	93
	4.1. What is Morally Special about Refugees?	93
	4.2. What We Owe to Refugees	97
	4.3. Distributing Refugees	109
	4.4. The Limits of Our Duties to Admit Refugees	114

Part II: Debating Immigration Policy in the Climate Context

5.	Facing a Heated Reality: Climate Change and Migration	121
	5.1. The Climatic Drivers of Migration	123
	5.2. The Case against Securitizing Climate Migrants	130
6.	Realizing the Conceptual Problems of Climate Migration	134
	6.1. The Fluidity of Categories	134
	6.2. The Issues of Causation	145
7.	Responsibilities toward Climate Migrants	153
	7.1. Realizing Obligations under Corrective Justice: Sidestepping the Issues of Causation	154
	7.2. The Responsibility to Forestall (Climate) Migration	166
	7.3. The Inappropriateness of Ex-Post Compensation and the Pitfalls of Arguing for Collective Resettlement	179
	7.4. Outlook: The Consequences of Irresponsibility	184
8.	Conclusion	187

Bibliography	197
Index	212

One

Introduction

Immigration has been a major driving force of recent developments in world politics. It took centre stage not only in Donald Trump's presidential campaign, but also in public British debates in the run-up to the vote on whether or not to leave the European Union. Likewise, in several European countries where right-wing political movements are on the rise, the issue of immigration is regularly found at the forefront of the political arena, with populists united among themselves on that issue.[1] Immigration is in the truest sense of the word a divisive issue. It divides societies politically, and it even threatens to tear them apart, as recent developments in the European Union foreshadow. This work is concerned with a phenomenon that could lead to a further dramatic increase of worldwide migratory pressure.

The phenomenon I will be concerned with is migration induced by anthropogenic climate change. Although projections vary greatly from about 50 million to 1 billion *further* migrants by 2050,[2] the most frequently found estimate is that of around 200 million further people on the move due to the direct or indirect effects of anthropogenic climate change.[3] It may be uncertain how many of these further migrants will move internationally (rather than nationally) and how many of them will be in a relevant sense *forced* to move.[4] But with the current number of worldwide migrants reaching an estimated 244 million in 2015, which according to the International Organization for Migration (IOM) is the highest number ever recorded,[5] and with the number of people forcibly displaced by war and persecution equally at a new all-time

---

[1] See Zakaria (2016, pp. 9–15) and Judis (2016, pp. 131–53).
[2] For an account of how these different estimates come about, see Piguet *et al.* (2011).
[3] See International Organization for Migration (IOM) (2016).
[4] The term 'migrant' is used here to cover all international non-tourist movement of people. Below in this introduction, I will define the different categories of migrants and refugees in some detail. In any event, I will keep quarrelling with such categories throughout later stages of my analysis.
[5] See International Organization for Migration (IOM) (2016).

high of 65 million in 2015,[6] the prospect of additional hundreds of millions clearly does not bode well. Sheer numbers will put under pressure all potentially involved sides: the affected people themselves, the destination countries, and the source countries.

How many people will eventually be induced to migrate because of climate change will depend — among other things — on the actual course that climate change takes. Here lies a major source of uncertainty.[7] Freak weather, droughts, and rising sea levels are likely consequences of a warming climate — and the less that is done in terms of mitigation of and adaptation to climate change, the more people these factors will induce to leave their homes.[8] Although much of this migration is estimated to take place within the home countries of the affected, many will have to cross international borders in their attempt to reach a safer place to live, and a disproportionate number of them will be from developing countries.[9] With a view to all these people who are induced to migrate because of climate change — be they voluntary or forced migrants — there exists a normative gap in international law. While there may be complementary protection under existing human rights law,[10] there are no tools designed specifically for dealing with the existence of climate migrants, let alone with their numbers. Nor is there much of an ethical debate on climate migration. These are two central recognitions that motivate the normative reasoning in this work. Morally-normative understanding is urgently needed so that policy makers can draw on it when finally tackling climate migration. *What do we morally owe to climate-induced migrants?* This is the guiding question

---

[6] See United Nations High Commissioner for Refugees (UNHCR) (2016). According to the UNHCR's terminology, of the 65 million "forcibly displaced" persons, 21.3 million persons were refugees (of which, in turn, 16.1 million were refugees under the UNHCR's mandate and 5.2 million Palestine refugees under the mandate of the United Nations Relief and Works Agency for Palestine), 40.8 million were internally displaced persons (IDPs), and 3.2 million were asylum seekers, i.e. those refugees still waiting for their asylum application to be decided and thus for their refugee status to be granted (UNHCR, 2016, pp. 2–3).
[7] For a sceptical stance on the issue of climate migration, see William Nordhaus (2013, pp. 102–14). In Nordhaus's view, "we know virtually nothing about the impact of global warming on future human migrations" (Nordhaus, 2013, p. 102). However, this appraisal is in contradiction to the larger part of the scientific literature on climate migration.
[8] See International Organization of Migration (2017).
[9] See Care *et al.* (2009).
[10] See Cournil (2011).

of my research. So far, this concern with climate migrants met with little attention in applied ethics and political philosophy.[11]

In the face of already staggering and still rising numbers of migrants worldwide, some moral theorists in current debates on the ethics of immigration keep debating about the necessity of a right to global free movement. By itself, this is not problematic. What is worrisome though is when those theorists wholeheartedly call for such a regime of global free movement to be instituted, in actual practice, that is. They defend a position which, if it were adopted, would risk leading straight into turmoil. Provided the demand for immigration will increase dramatically due to climate change, there are currently no indications and few (if any) reasons for hope that available entrance places will rise in proportion with that increase. In this context where numbers are indeed high enough to overwhelm societies, calls for open borders are neither realistic nor especially appealing. And yet, more restrictive immigration policies are hardly the better alternative.

It appears that reasoning on the phenomenon of migration—and particularly of climate migration—leads to a tension which is the result of two hardly reconcilable views. On the one hand, high levels of immigration as associated with future climate-related movement mean *potentially* high costs for democratic countries so that these *countries should have a right to exclude* migrants when the costs become too high. At least, they appear to have particularly good reasons for exclusion when the number of migrants rises to unprecedented levels. Indeed, even many of those philosophers who argue in support of open borders concede that free movement is a reasonable (and feasible) option only in a just world in which many people will indeed not feel the need to migrate in the first place. On the other hand, it is far from clear whether those countries also have this *right to exclude* when the presumably *special* group of climate-induced migrants is concerned. One could assume that they do not. For is it not the case that, in so far as those countries are and have been emitting greenhouse gases, they have contributed causally to the existence of climate migrants in the first place? To the extent that this group of climate migrants now threatens to be the numerically largest one of all, and to the extent that the right to exclude does not apply to precisely this future largest group (as one could presume), then in what sense could we meaningfully speak of a right to exclude at all? Against the backdrop of this tension, I will point

---

[11] For two notable exceptions, see the articles by Wyman (2012) and Penz (2010). A utilitarian approach to the question of what we owe to climate migrants is provided in an article by Nawrotzki (2014).

out how the phenomenon of climate migration challenges *both* positions generally taken in the current debate: that of the defenders of the right to exclude and that of the proponents of a right to global free movement.

In addition to this fundamental tension, there is a series of analytical difficulties that obstructs meaningful debate on what we owe to climate migrants. First, there are the issues of causation: if it is often not possible to tell with certainty whether and to what extent a particular migrant is indeed a climate migrant — because the causal relationships between emissions of GHG and migratory processes remain essentially blurred as regards the specific case — then how can one meaningfully speak of "climate migrants" in this causally blurred context? And how could one make moral and practical sense then of the intuition that climate migrants are *special* and somehow deserving of "our" special moral attention? And in so far as the climate — or the environment — is often only one among several other factors that *together* cause migration, then speaking of "climate migrants" would appear to become more problematic still. The second analytical problem concerns the *categorization* of climate migrants. I will defend the thesis that climate migration, more than other forms of migration, defies the distinction between "voluntary" and "forced" movement, which is after all a distinction that informs the current moral debate on immigration in so far as (from a human rights perspective) forced migrants are considered morally entitled to protection while voluntary "normal" migrants are not. Now, in what sense does a person who "decides" to leave her drowning island state *long before* the water is literally up to her knees really and freely decide to move? Or is this — despite her house still standing and still being dry — already an instance of forced migration? What is lacking here is a solid conceptual understanding of voluntariness and forcedness in the particular case of climate migration. Such conceptual understanding is after all a precondition for debating our moral obligations toward climate migrants. Therefore, the just mentioned analytical difficulties will be at the forefront of the ensuing debate. This conceptual work will prepare the ground for all subsequent reasoning on what we owe to climate migrants.

After this rough overview of the challenges that the following analysis will deal with, it is worth pointing out how this work's argumentation proceeds chapter by chapter. Now, placed in front of this work's main part, there will first be *three further sub-chapters to this introduction*. In the first such sub-chapter, I will make a series of important methodological remarks on the kind of perspective that I will take in the ensuing normative debate on migration and climate change. In the second sub-chapter, I will be concerned with terminology, and more specifically with a first approximation to the distinction between (non-refugee) migrants, refugees, and asylum seekers. Finally, in the third sub-chapter to this introduction, I will sketch along rather general lines the economic and social effects of migration, and particularly of immigration. Subsequent chapters will build on this empirical understanding and on the awareness it raises, namely that certain forms of immigration can come with costs for all involved sides. Its major task (i.e. its function in the overall argumentation of this work) is to make plausible the modest hypothesis that — all *chances* of immigration notwithstanding — *certain* forms of immigration mean a *risk* to the host country's societal institutions.

The second chapter will be fundamental in the truest sense of the word. It serves the purpose of establishing the moral principles that will then be applied to the related issues of climate change and migration. In particular, I will set forth a conception of basic human rights, which will then anchor all moral reasoning in this work. My elaborations on basic rights are guided by the practical concern that people find these rights ultimately realized and secured, globally. For that purpose, I will propose an account of how states should divide among themselves the task of realizing their respective citizens' rights, and I will set forth a conception of a related principle which becomes important once people find their basic rights invaded: the normative principle of corrective justice.

In chapter 3, that concern with basic rights will be applied to the reality of immigration. It elaborates in some detail on the question of whether there is a *right to exclude*. After rejecting the view that there is a (human) right to global free movement, I will contend that the exclusion of migrants from the political community of a country is in principle consistent with the assumption of all people's equal rights. Going beyond this assessment that exclusion is in principle consistent with equal rights, I will argue that states have special duties toward their own citizens. The right to exclude thus figures in my account as an indispensable authority that states should have in order to protect certain conditions which allow a democratic society to function in the first place and to protect such special duties. With its emphasis on the importance of community, this assessment resonates with existing strands in the philosophical literature on immigration. But rather than merely repeating conventional thinking, the way in which I draw on the mutual dependence of community and human rights to bring home my case presents a new approach. The systematic function of this chapter is to set the stage for an upcoming concern though. It is precisely by establishing the view that exclusion is an important or even indispensable authority that the deeper concern with the looming reality of climate migration will become visible. Potential host countries have on their side powerful reasons for the exclusion of migrants—but these reasons will have to be pitted against whatever special claims to admission climate migrants could bring forward.

Chapter 4 is about the case of *refugees*. It is pointed out why refugees are a special case, why we have particularly strong obligations toward refugees, and how refugees should be distributed among the world's states. While this chapter has the systematic function of providing a reference point for later debate on what is owed to *forced* climate migrants, it can also be read as a relatively free-standing and independent argumentation. Responding to shortcomings in the current normative debate on refugees, this chapter will include a discussion of how our obligations toward refugees are *limited*. These assessments will inform later debates on the limits of our obligations toward forced climate migrants.

It is in chapter 5 that the phenomenon of climate migration, while indicated at several points in earlier chapters, is finally and systematically introduced. Before climate migration can become the object of normative analysis, it is once more essential to gain a solid empirical understanding: this time of the distinct ways in which climate change will induce people to migrate. But this is not the only purpose of this fifth chapter. It will also refer to, and examine critically, the tendency in academic and policy circles to treat climate migration as a security issue or, put in technical terms, to *securitize* the issue. This security discourse is rooted in the fear that the sheer quantity of future climate migrants will overwhelm potential destination countries. Irrespective of whether or not this fear is justified, the problem with such securitization discourse is that it tends to degrade the affected people to mere objects of a security concern. This observation then leads me to a sharp critique of this way of framing climate migration.

Departing from the empirical insights gained in chapter 5, I will point out in chapter 6 why climate migration is such a difficult phenomenon to come to terms with morally. There are two conceptual problems already indicated above. First, there is the problem that it is difficult to draw a meaningful line between instances of voluntary climate migration and instances of forced climate migration. While this is a problem with a view to most migratory processes, I will explain why it is especially accentuated in the case of climate migration and how this circumstance impedes moral investigation into the question of what we owe to climate migrants. This discussion will inform my approach to categorizing climate migrants. The second set of problems is referred to as the "issues of causation". As I will point out, normative debate on what we owe to climate migrants must deal with the additional analytical hurdles that it can hardly be said with certainty *whether* and *to what extent* a climate migrant was indeed induced to migrate because of the climate, that even if a "harmful" link between climate change and migration could be established, it would remain essentially unclear to which country this presumed harm can be attributed, and that the climate is often only one among a multitude of other social, political, or economic factors that combine for migratory pressure.

In chapter 7 I will then bring forward a substantive argument for certain responsibilities we have toward climate migrants. I will substantiate the claim that those proposed responsibilities can *at least in principle* be grounded in the normative principle of corrective justice. Such responsibility under corrective justice would then *complement* whatever more general human rights-related obligations there may be. This application of corrective justice will require sidestepping the just mentioned analytical hurdles (i.e. the "issues of causation"). If such hurdles can indeed be sidestepped and if corrective justice is indeed applicable, then an important contribution to the current debate on climate migration will be made. For in this debate it is widely assumed that a case for responsibility under corrective justice cannot be maintained in light of the blurry causal relationship between emitters and climate migrants. My assessment in that chapter will then allow me to make concrete proposals on how our responsibilities toward climate migrants could be institutionalized. At the end of this chapter I will also discuss what happens if the proposed responsibilities are not fulfilled by any or most of the world's potential helping countries. That discussion will underline that irresponsibility has a price, for all sides, and that one had better fulfil one's responsibility in the first place.

1.1. Methodology: Debating Immigration in an Imperfect World

In a recent book on the political philosophy of immigration, political philosopher David Miller casts serious methodological doubt on the reasonableness of applying a strong cosmopolitan perspective to the issues of migration, immigration, and refugee policy. His contention is that if prior to the debate on immigration one already holds that the current international order is unjust, that rich countries should distribute a much larger share of their wealth to poorer countries, and that a system of territorially confined states with borders is fundamentally unjust, then it would be unclear in what ways one could meaningfully and constructively discuss the real problems of immigration. For after all, they exist *only* in our imperfect present world which is characterized by precisely these presumably unjust features:

> If we assume a fully just world, there would be no refugees, and no one seeking to escape desperate poverty. So all the factors that make immigration such a controversial issue for us would be absent in this hypothetical world. One could therefore "solve" the immigration question by prescribing that the world should become stateless or distributively just, but how much practical light would that throw on our own predicament?[12]

This suggests that at least to some extent one must take for granted those preconditions and circumstances of the real world that make immigration a controversial policy issue in the first place. In an imperfect world, immigration can compromise the functioning of established societal institutions of the host country: especially when the pace of change through immigration is too fast, citizens could feel culturally alienated and turn their back to social and political life;[13] or levels of social trust could be reduced so that there would be a trade-off between high immigration levels and a system of social welfare that depends on those levels of social trust. These are just two scenarios, and while empirical evidence will be provided for such theses in subsequent chapters, the methodological point to be made here is that it makes sense to accept and work with available evidence on the way in which real people react to immigration.

One could of course dismiss as narrow-minded and intolerant that people, depending on what kind of and how many strangers immigrate, feel culturally alienated. But then again, such a move would just mean to assume away the very reason why a normative debate on immigration is needed in the first place. If all people were sufficiently open-minded, if all members of a political community warm-heartedly welcomed people from other cultures, and if all people generously accepted economic disadvantages associated with the arrival of large groups of foreign people, then maybe immigration would pose no challenge, but then again there would be no problem worth discussing in the first place. Therefore, one should take for granted at least to some extent the attitudes and reactions of real people toward immigration.[14]

---

[12] Miller (2016, p. 17).
[13] See Zakaria (2016, p. 15).
[14] The approach proposed in this section is inspired by David Miller's "political approach" as he characterizes it in Miller (2016, pp. 16–19).

What this means methodologically, i.e. for the way in which the issue of migration is debated in this work, is that I will take what David Miller calls a *political approach* as opposed to what I will refer to as an (idealistic) "individualist-cosmopolitan approach".[15] The difference between the two approaches as here understood concerns not only the degree of realism that informs the analysis. They also differ centrally in so far as they address their claims to different addressees: the individualist-cosmopolitan approach does not take the constraints of reality too seriously (at least less so than the political approach), and the addressees of the moral demands formulated under that individualist-cosmopolitan framework often appear to be first and foremost individuals. Such an approach would require and presuppose that individuals act in a certain way that is in line with certain moral principles that inform analysis. As concerns the challenges related with immigration, one could then ask the citizens of host countries, who currently tend to complain a lot about the reality of immigration, to be just a little bit more open-minded and tolerant.

However, to assume and expect that they will actually act in accordance with those demands, i.e. that they alter their behaviour "for the better", would mean once again to assume away precisely those conditions of reality that produce the real-world challenge of immigration in the first place. If one applied this individualist-cosmopolitan approach to, say, the specific challenge that social trust among citizens and thus the functionality of social systems tend to be strained through immigration, then one could simply criticize the individuals' reaction to immigration. One might simply demand of them that they not trust less in others only because their society gets more culturally heterogeneous. And indeed, this move would nicely eliminate one of the major problems societies face when immigration takes place: "So any problems that immigration may currently pose for the survival of the welfare state can be resolved by promulgating a norm that people ought to follow, one that follows from basic moral principles."[16] Such an approach to the morality of immigration will yield results that are of little practical use though. They will, presumably, be little helpful as a guide for policy makers. Worse still, for policy makers to draw on such an approach exclusively and use it as the sole guide of their rhetoric would expose them as utterly unworldly.

---

[15] Note that I diverge here from Miller's terminology. Miller simply refers to the opposite of the described "political approach" as the "ethical approach".
[16] Miller (2016, p. 18).

The political approach, on the other hand, takes a much more realistic view of the way people behave in response to immigration. Rather than with individuals, the normative proposals a *political approach* aims to formulate are concerned with the design of policies and institutions, and those proposals are thusly directed, at least primarily, at political actors in a position to influence and change those policies and institutions. This requires that political, sociological, economic, and cultural realities are taken into account when such morally-normative proposals are developed. Moreover, it will be necessary to take for granted a certain (indefensible) degree of unjust inequality in the world, and it will be taken for granted that the people act toward immigration in certain ways the philosopher may in principle find inappropriate. But rather than "simply telling people that they should be less prejudiced and more trusting of strangers",[17] it is more promising to work with people's real attitudes, accept them by and large as given and then set out to find practical moral approaches to the imperfect reality we face.[18]

The precondition for taking such a political approach is that one draws on the best empirical understanding that is available on the issues of migration, immigration, and climate change. This understanding, especially of the way the social world works, will then have to be considered when one tries to assign concrete duties to concrete actors. The implication of such a political approach is not only the possible reward of more practically useful results, but also an additional difficulty for the theorist. As Henry Shue observes:

> Rights theory becomes much more difficult, because one cannot offer only conceptual analyses. The conceptual analyses, which of course remain essential and perhaps fundamental, must be given in

---

[17] Miller (2016, p. 18).

[18] Note as an important caveat that such a perspective is convincing only when taken by an uninvolved party, for example a normative theorist who with some distance tries to make recommendations on immigration policy. However, it would be less convincing for an involved party to draw on that perspective. For example, it would be rather doubtful for the citizens of a potential host country (i.e. an involved party) to say that they will react to the immigration of a certain group of migrants in a certain problematic way, and that it would therefore be problematic to let that group immigrate. They could in this way release *themselves* from any responsibility there otherwise would have been toward that group of migrants (see Brezger, 2016, p. 67). In contrast, the moral philosopher debating immigration policy will enter the scene as an uninvolved party that is authorized to take for granted, at least to some extent, the attitudes of involved parties. Likewise, governments are, presumably, often well-advised to take the same perspective when making decisions on immigration policy.

conjunction with an analogue of what the social scientists call "operationalization". One must spell out, at least a little bit, what it would actually mean for a certain right to be fulfilled and enjoyed. This entails analyzing which tasks must be performed [...] and which kinds of people can reasonably be expected to perform them [...].[19]

Finally, it is worth noting that the political approach as just proposed has some relevant parallels to Max Weber's conception of an "ethic of responsibility", which he famously opposes to an "ethic of ultimate ends":[20]

> You may demonstrate to a convinced syndicalist, believing in an ethic of ultimate ends, that his action will result in increasing the opportunities of reaction, in increasing the oppression of his class, and obstructing its ascent—and you will not make the slightest impression upon him. If an action of good intent leads to bad results, then, in the actor's eyes, not he but the world, or the stupidity of other men, or God's will who made them thus, is responsible for the evil. However a man who believes in an ethic of responsibility takes account of precisely the average deficiencies of people; as Fichte has correctly said, he does not even have the right to presuppose their goodness and perfection.[21]

The political approach with its explicit focus on institutional design as opposed to individual conduct goes beyond Weber's conception of an "ethic of responsibility": but the decisive point they have in common is the weight they give to actual consequences. Indeed, none of the normative proposals in the following chapters will be made in the spirit of someone whose conduct is guided by what Weber identifies as an "ethic of ultimate ends": a proposed immigration-related measure that leads to unacceptable results is invariably considered a bad measure, no matter what benevolent intentions may motivate that proposal and no matter what moral principles it is derived from. Instead I will look at the possible consequences of the measures proposed, and more precisely at how such measures bear on the equal rights of all people. However, as there will invariably remain some degree of uncertainty as to what the actual consequences of a given measure will be, the political approach will centrally rely on the language of *risks*. As part of the political approach, a risk-ethical perspective will thus be built into the overall rights-based framework presented in the following chapter.

Finally and closely related to that concern with risks (which of course remains to be substantiated), another central concern in this

---

[19] Shue (2004, p. 226).
[20] Someone who is guided by the maxims of such an idealistic "ethic of ultimate ends" will not give much weight to the *actual* consequences of his actions.
[21] See Weber (2004, p. 47).

work's political approach will be with incentives. When assigning concrete duties, the main concern should always be that the right-bearers actually and eventually come to enjoy those rights, and this will make it necessary to be sensitive to incentive structures, which therefore will be integrated into this work's normative framework as well.

In sum, the normative framework to be applied to the related issues of immigration and then climate-induced migration will be this: a realistic rights-based theory that draws where needed on a risk-ethical perspective and which is, after all, sensitive to the importance of incentive structure. This leaves us with a dynamic framework, and it will inform the political approach that is to guide the normative analysis to come.

1.2. Terminology: Refugees, Asylum Seekers, and Other Migrants

Countries categorize migrants and give them corresponding labels. Depending on what category a migrant finds herself in, the country may then deem that migrant to be more or less deserving of priority for entrance. I will now propose and define three migrant groups, namely refugees, asylum seekers, and non-refugee migrants, like economic migrants. This rough grouping is meant to orient later attempts to come to terms with the "new" category of "climate migrants", if such a category exists.[22]

Refugees

The most frequently referred to and legally most authoritative definition of a refugee is the one established by the 1951 Geneva Convention and its 1967 protocol.[23] In the Convention, which was signed by 145 state parties including all liberal democratic states,[24] a refugee is defined as a person who

> owing to a well-founded fear of being persecuted for reasons of race, religion, nationality, membership of a particular social group or political

---

[22] For example, one further category of migrants that could be distinguished from the three groups proposed here is social migration, covering family channels (family reunification) and channels for those who have ethnic ties and migrate to return home. However, this group is of no systematic interest in this work so I will largely neglect it. For more on social migration, see Goldin *et al.* (2011, pp. 140–47).

[23] For a short introduction to the international law of refugee protection and for helpful further references see Goodwin-Gill (2014); and for a more thorough account, see Betts and Loescher (2010).

[24] UNHCR (2015, pp. 1–5).

opinion, is outside the country of his nationality and is unable or, owing to such fear, is unwilling to avail himself of the protection of that country; or who, not having a nationality and being outside the country of his former habitual residence, is unable or, owing to such fear, is unwilling to return to it.[25]

The Convention was developed in a specific historical context after the Second World War and early during the Cold War with the ongoing experience of oppressive regimes. As an understandable result, its refugee definition focuses on persecution as one of three crucial features: for an individual to be considered a refugee it is necessary that she is or has a justified fear of being persecuted. As a second feature, this persecution must take place on account of one (or more) of the five grounds of race, religion, nationality, social group, and political opinion. The third main feature specifies that the individual has to be outside her country of nationality. Now, could one simply bend this definition of refugeehood in such a way that forced climate migrants would be considered refugees? Most authors that I draw on in this work agree that this is hardly a convincing way of dealing with climate migrants. The latter differ relevantly from Convention refugees in so far as—all possible parallels with flight from armed conflict notwithstanding—it is not clear in what sense climate migrants could be regarded as persecuted.[26] What further complicates the applicability of the Convention is that only part of the world's (forced) climate migrants will be outside their respective country of nationality.[27] It appears then that the Geneva Convention is insufficient as a tool for coping with the challenge of climate migration.[28]

In diverging from the overly narrow Convention definition, I will employ in my analysis a broader conception of refugeehood. This is Matthew Gibney's definition of refugees, which at least provisionally I will rely on as a working definition:

> [P]eople in need of a new state of residence, either temporarily or permanently, because if forced to return home or remain where they are they would—as a result of either the brutality or inadequacy of their state—be persecuted or seriously jeopardise their physical security or vital subsistence needs.[29]

---

[25] See UNHCR (1951, 1967).
[26] See Docherty and Giannini (2009, pp. 357-59), Lyster (2015, pp. 433-38), and McAdam (2011, p. 38).
[27] See Warner *et al.* (2013, p. 20).
[28] Lyster (2015, p. 38).
[29] Gibney (2004, p. 7).

By this definition, what matters is that the affected person is in real danger or exposed to serious risks. It does not matter whether the source of that danger or risk is a state authority. The view that this broader focus is the morally more appropriate one is a widely shared view in discussions on the morality of immigration.[30] By Gibney's definition refugees need not be outside the territory of their state of nationality (although they need a new state), which is an important difference from the Geneva Convention's definition. A further strength of this definition is that it adds to the narrower Geneva Convention's focus on persecution (by a brutal state) the possibility of a state being *inadequate* to live up to the responsibilities it has towards its subjects. When needed, I will henceforth distinguish between the broader term "refugee" and the narrow term "Convention refugee". The broader one will dominate the following discussions though.

Asylum Seekers

There is a further more technical distinction that can be drawn between refugees and a sub-group of asylum seekers. Technically speaking, refugees normally flee to neighbouring countries and to refugee camps near the border, and in most cases they never go through an individual determination procedure before they are formally recognized as refugees. Asylum seekers, on the other hand, who are by far the minority of worldwide refugees, apply for humanitarian protection— often in distant countries.[31] On their route to reach those countries, many cross international borders illegally. Once they arrive, they claim refugee status, and the distinctiveness of the asylum seeker is then that her status has not yet been determined by a given state's determination procedure.[32] Some asylum seekers will turn out to be refugees, but others will not.[33] It may seem as if this distinction between refugees and asylum seekers sheds light on a "merely" procedural issue then, for even before an asylum seeker is actually granted refugee status he is or is not—*objectively*—a refugee in the morally relevant sense. This point is also clarified by the UNHCR, stating that the "formal recognition of someone, for instance through individual refugee status determination, does not establish refugee status, but confirms it."[34]

---

[30] See, for example, Shacknove (1985) and Carens (2003, p. 95). The potential danger of broadening the refugee definition too much is that, at some point, an impracticably large number of people in the world would count as refugees.
[31] See Goldin *et al.* (2011, pp. 147–49).
[32] See Castles *et al.* (2014, p. 222).
[33] Goldin *et al.* (2011, p. 152).
[34] UNHCR (2008, p. 41); similarly quoted in Goldin *et al.* (2011, p. 148).

There is, however, also a more clearly *moral* issue that the distinction between asylum seeker and refugee makes visible. Assume the perspective of a particular receiving state: of all the world's refugees, only a small portion shows up — as asylum seekers — at the borders of that particular country (whereas other refugees remain somewhere else, say in refugee camps, while yet others show up at other countries' borders). Assume further that all of them *are* objectively refugees. The moral issue is then this: why should the destination country help those refugees who arrived as asylum seekers rather than all the others who have not reached that country, who are still on their dangerous journeys, who are still in far-away refugee camps or who remain within their own country where they are not safe? Many of them may turn out to be even needier than the few (lucky, rich, or strong enough) who reached the destination state's safe haven. To say that the destination state has toward those refugees (i.e. asylum seekers) at its borders a stronger obligation to assist than toward those refugees who remain distant strangers would appear to be a stance that attaches just too much relevance to the arbitrary circumstance of geographical proximity.[35] This is an issue I shall come back to later. For now, it suffices to acknowledge that some challenging immigration-related questions come into full view only when a conceptual distinction between refugees and asylum seekers has been drawn.

(Non-Refugee) Migrants

The final category I shall provisionally deploy here tries to capture the large group of people who are not refugees in the described sense, and who can thusly be subsumed under the category of "non-refugee migrants". This group of non-refugee (economic) migrants is a most heterogeneous one. It covers the wide range between the following two extremes: at the one extreme high-skilled migrants who move (often between developed countries) to improve an already decent economic situation, and at the other extreme those migrants whose quality of life in their home countries is far from decent and who therefore leave their countries in search of an economic situation that is at least less precarious. In other words, the category of economic migrants may encompass very different instances of migration: a businessman moving from Germany to Canada and an Ethiopian farmer who leaves Ethiopia for some other country in order to escape poverty. It goes without saying that the more interesting group to deal with when concerned with the morality of immigration is the latter one, in part

---

[35] See Gibney (2004, p. 10).

because the high-skilled German businessman faces fewer constraints when migrating to other developed countries than the low-skilled Ethiopian farmer would, and because so much more seems to be at stake in the case of people migrating to escape poverty.

Now, a central problem seems to be that some economic migrants' motives to leave come close to the considerations that were introduced above to denote the category of refugees. If a refugee is defined as a person who flees in order to escape conditions that make a decent life impossible or that are (or which she perceives to be) life-threatening, then the claim of at least some economic migrants will vary only slightly from those of refugees. And yet, to ask for admission into another country on the grounds of dire poverty at home is a slightly less urgent claim than to say that one faces a life-threatening situation at home. On the continuum between voluntary and forced migration, the difference between refugees and some non-refugee migrants is indeed only a gradual one: although even refugees will always retain a rest of agency that would render it problematic to speak of them as having literally no choice but to move, economic migrants retain a higher degree of agency and voluntariness. With the difference between these two categories of economic migrants and refugees being only a gradual one, one should certainly be careful in drawing it. And yet, in a world with only scarce entrance places for the world's migrants, this morally relevant distinction between voluntary and forced migration must be drawn somewhere. In a world where states define limits to their capacity and thus to their obligation to take in needy outsiders, "it makes sense to prioritise claimants for entrance; and in a conflict between the needs of refugees and those of economic migrants, refugees have the strongest claim to our attention."[36] It remains to be seen how helpful this verdict is when, later in this work, it comes to the case of climate migrants.

1.3. Setting the Empirical Stage: The Economic and Social Effects of Immigration

It is indispensable to have a basic understanding of the effects of immigration established here so that later normative debate can be built on that understanding. What would the effects of more open borders be? This question shall structure the following account of the consequences of migration. The thesis—or rather the claim to be substantiated—is that certain forms of immigration *potentially* put a strain on relevant institutions of the host society. In so far as this is indeed a

---

[36] Gibney (2004, p. 12).

rather modest claim, it does not appear to be necessary to provide an overly lengthy examination of the current empirical literature on the effects of immigration. Instead, it will suffice to point to some general *tendencies* and observations regarding the effects of immigration. Clearly, there is some social-scientific controversy regarding the specific effects of specific instances of immigration. However, after the following short account with its rough overview of such social-scientific accounts, there will be one thing that can be said with certainty, namely that at least certain forms of immigration come with costs for the host society. And these costs are sometimes most appropriately couched as risks.

Economic Effects

Economic orthodoxy tells us that open borders will have positive net effects globally. If people can move freely, they can leave countries where the demand for labour and productivity are low. Leaving behind these places where people are not needed and where they are therefore often unable to make a decent living, they would move to other countries where the supply of labour is scarce, where they are more productive and where they are more likely to become economically better off.[37] This would boil down to a more efficient global allocation of resources, which in theory promises welfare gains.[38] And indeed, the free circulation of people across borders appears to be a demand of economic globalization. In its process, states and their national economic policies have become bound by the reality of the increasingly free and cross-border circulation of goods, capital, services, and information. But while states accepted reduced control over the just mentioned productive factors, they retain it where the movement of people is concerned.[39] This asymmetry is problematic in so far as economic turmoil in one country (e.g. inflation or demand shocks) could be imported by other (little diversified) countries which depend on the country in turmoil.[40] Against such dependencies open borders would have the effect of a safety valve: if in one country the demand for labour declines unexpectedly (due to globalized economic processes beyond that country's control), newly unemployed parts of the workforce could leave the country and thusly counter increasing unemployment.

---

[37] See Kukathas (2005).
[38] See McDowell *et al.* (2006, p. 601).
[39] See Goodin (1992, pp. 12–13), Truong and Gasper (2011, p. 9), and Bader (1997, p. 1).
[40] See McDowell *et al.* (2006, pp. 875–80).

An important aspect related to the net gains to be expected from open borders, or at least from less restrictive border controls, concerns the remittances of migrants. Many migrants transfer large parts of the money earned in their host country back to their country of origin. For some, the idea of improving the situation of those left behind through remittances is the reason to migrate in the first place.[41] The money flowing into developing countries through remittances is around three times higher than what developed countries spend on official development assistance (ODA). By 2014, global remittances to developing countries alone have reached a total of $441 billion.[42] In some countries the income through received remittances constitutes around a fifth of GDP (gross domestic product).[43] Given these numbers, the positive effects of (more) open borders especially for developing countries seem potentially high. In some cases, however, the gains from remittances could be offset by the losses especially poor countries incur when talented and well-trained (young) people leave the country, a problem commonly referred to as "brain drain".[44] What is more, it is not necessarily the case that remittances always reach the poorest people in developing countries—for it is not the most destitute who can afford to have family members migrate.[45] In a world of open borders, however, even the poorest would probably find it easier to migrate.

Finally, I come to the economic consequences of immigration—especially of large-scale immigration—on destination countries. As concerns wages, it is a general truth that immigration produces losers and winners, at least when short time frames are considered. The immigrants tend to win (especially where they come from extremely poor countries), the more skilled workers of the host society also tend to win, and the poorer and less skilled workers tend to lose. Their wages generally drop considerably as it is them who compete most directly and immediately with the newly arrived migrants.[46] However, these assertions on the development of wages depend on who immigrates (or who is admitted). Skilled immigrants, for example, would complement

---

[41] See Bodvarsson and Van den Berg (2013, p. 212).
[42] See World Bank (2016, p. 20).
[43] See World Bank (2014).
[44] For an insightful discussion of whether "brain drain" can, from a moral standpoint, justify immigration restrictions, see Brock and Blake (2015) and Oberman (2013).
[45] See Bodvarsson and Van den Berg (2013, pp. 197–99); or Pogge (2002, pp. 112-13).
[46] See Borjas (1999, p. 13).

rather than compete with the unskilled workforce and thus raise their productivity.[47]

Once one moves beyond the specific concern with wages to a more general concern with labour markets, it becomes ever clearer that large-scale immigration comes with a high price at least for some groups of the indigenous population. Let us focus on the group (or class) of poor workers. They are negatively affected by mass immigration of low-skilled workers in so far as labour markets are flooded with further supply of workforce, which drives down wages and potentially leaves some people newly unemployed.[48] The notion that any immigration policy will produce winners and losers can thus be concretized: the losers of large-scale immigration are often the working-class poor, many of which are themselves previous immigrants. As George Borjas notes: "Immigration [...] generates a sizable redistribution of wealth in the economy, reducing the incomes of natives who are now competing with immigrant workers in the labor market and increasing the incomes of capitalists and other users of immigrant services."[49]

In sum, it appears to be disproportionately the relatively wealthy members of society who benefit from whatever gains through migration there may be. In theory, it would of course be an option to redistribute the gains to the poor, and it stands to reason that their (or even society's) acceptance of immigration depends on such redistribution in order to make immigration palatable to all. Unfortunately, as current research suggests, certain forms of immigration tend to reduce the willingness of existing citizens to make such redistributions in the first place.[50] This is the concern I will turn to now.

Social Effects

> Immigration is the final frontier of globalization. It is the most intrusive and disruptive because as a result of it, people are dealing not with objects or abstractions; instead, they come face-to-face with other human beings, ones who look, sound, and feel different. And this can give rise to fear, racism, and xenophobia. But not all the reaction is noxious. It must be recognized that the pace of change can move too fast for society to digest.[51]

Even though immigration is widely understood to have beneficial social effects in so far as diversity (through immigration) tends to foster

---

[47] See Collier (2013, p. 113).
[48] See Dustman *et al.* (2008, 2013), and Borjas (1995).
[49] Borjas (1995, p. 18); and for the same evidence see Ottaviano and Peri (2012).
[50] See Collier (2013, p. 113).
[51] Zakaria (2016, p. 15).

a more dynamic, creative, and cosmopolitan society, there are social costs faced especially by communities that experience rapid and large-scale immigration.[52] Such costs can be high and they are related with the often observed *tendency* that *social trust* and a sense of community are deteriorated when immigration occurs too quickly and too extensively. It is this tendency of a reduction in *mutual regard* and *social trust* among citizens that I will be centrally concerned with in the first parts of this discussion of the social effects of immigration. So before I turn to empirical evidence, I will in a first step provide a conceptual understanding of the meaning and relationship of mutual regard, social trust, and levels of inner-societal redistribution.

Mutual regard denotes a feeling of trust and sympathy toward co-citizens that, presumably, allows for an attitude of loyalty and solidarity toward them. The idea is that a certain level of mutual regard and social trust fosters the willingness to cooperate with co-citizens in ways that make elaborate schemes of social welfare possible. So the presumption is that where this sense of community exists within a given society, the possibility of successful social cooperation is facilitated, the willingness to make financial transfers to needy members of the society is fostered.[53] As Kenneth Newton sums it up:

> Trust makes it possible to maintain peaceful and stable social relations that are the basis for collective behaviour and productive cooperation. Trust involves risks, it is true, but it also helps to convert the Hobbesian state of nature from something that is nasty, brutish, and short, into something that is more pleasant, more efficient, and altogether more peaceful. Social life without trust would be intolerable and, most likely, quite impossible.[54]

Now, as some authors in the social scientific discussion on the social effects of immigration suggest, levels of mutual regard and social trust in a society tend to decrease when immigration occurs at a certain pace and on a certain scale.[55] This view is not uncontested and I will put it into perspective in the course of this section. But it bears looking more closely here at those findings which suggest such a negative correlation between immigration and levels of social trust and which lead Robert

---

[52] See Goldin *et al.* (2012, p. 173).
[53] See Coleman (1988, pp. 100–05); and for a more systematic account of the role and relevance of social trust (or social capital) for the functioning of societal institutions, see Ostrom (1990) and Fukuyama (1995). Fukuyama examines why (and points out that) economic prosperity requires a certain degree of trust among the participants of the economy.
[54] Newton (2001, p. 202).
[55] See Putnam (2007).

Putnam to state a "tradeoff between diversity [through immigration] and community"[56] in the short run. Similar to Putnam, Alesina and Glaeser find that social welfare spending tends to correlate negatively with the level of cultural diversity within society.[57] The provocative and controversial claim to be distilled from such evidence could go as follows: the more immigrants there are, and the more culturally distant these immigrants are, the more difficult it becomes to maintain demanding social institutions. Pointing to what he claims to be an increasing body of empirical evidence, Collier concludes that "[a]s predicted by the theory, the greater the level of cultural diversity, the worse the provision of redistributive public goods."[58] That such claims rest on a solid enough empirical foundation is suggested by a meta-analysis carried out by Miller and Ali: examining the research so far available on that subject, they find general support for the thesis that there exists a correlation between immigration, levels of interpersonal trust among citizens, and levels of social spending.[59] According to Miller, "most social scientists" participating in current debates hold this view.[60]

So to the extent that such empirical contentions are resilient, they raise a relevant ethical concern: if immigration ultimately leads to less financial transfers from wealthier to poorer members in society, then—all else being equal—inner-societal levels of inequality will tend to rise.[61] I can claim no competency for assessing such empirical evidence, but there appears to be a sufficiently large group of authors who do warn of the social effects of certain forms of immigration. And it clearly has some commonsensical appeal that at least high rates of immigration will *at some stage* have detrimental effects on the institutions of the host society.[62] The least the above mentioned findings seem to allow for then is the balanced verdict that while "moderate migration is liable to confer overall social benefits, [...] sustained rapid migration would risk substantial costs".[63] This, in turn, suggests that regarding the possibility of detrimental social effects on society through immigration, a risk-ethical perspective is indispensable when debating the morality of immigration.

---

[56] Putnam (2007, p. 164).
[57] See Alesina and Glaeser (2005, p. 141).
[58] Collier (2013, p. 85).
[59] See Miller and Ali (2014).
[60] See Miller (2016, p. 64).
[61] See Collier (2013, pp. 83–85).
[62] See Zakaria (2016, p. 15).
[63] Collier (2013, p. 63).

Now, while there may remain much controversy and uncertainty on the relationship between immigration and the possibility of maintaining demanding institutions of social welfare, there appears to be some agreement on the following more modest claims. First, it can be observed that in countries experiencing a sudden growth in minority populations the level of social spending rises at least more slowly.[64] And second, whatever negative correlation between immigration and social spending there may be, most detrimental effects can in principle be countered by determined measures in the fields of, for example, integration and social policy.[65] "Countries that have adopted the most comprehensive multiculturalism policies [...] have been more successful at maintaining elements of a welfare state than those countries that have resisted such policies."[66] But even such a more optimistic appraisal that emphasizes the possibility of offsetting policies ultimately confirms the general notion that the overall number of immigrants entering the country is of significance and that immigration is potentially costly at least in the short run.

After all, and on pain of repeating this decisive point, the empirical findings drawn on in this section all support at least one rather modest claim: that with the society getting more diverse through (rapid) immigration, there is at the very least an increased *risk* that existing social institutions erode and become more fragile. According to Paul Collier, social cooperation games that underlie complex societal institutions are only *locally stable*: if social trust is reduced "beyond some unknowable point", the host society risks surpassing a threshold at which cooperation games become fragile.[67] What the host society risks then is the collapse of such cooperation games and therewith the functionality of those institutions that depend on cooperation. The recognition of this mere local stability of social cooperation games and the uncertainty regarding the point at which cooperation becomes unstable provides us with a further reason for assuming, wherever needed, a risk-ethical perspective when debating immigration.

In concluding this chapter's overview of the economic and social effects of immigration, it may be said that the modest thesis that certain forms of immigration potentially put a strain on certain societal institutions clearly found some substantial support in the adduced social scientific findings. Although the effects of immigration as outlined here can only insufficiently sketch what is in actual fact a wide

---

[64] See Soroka *et al.* (2006, p. 278) and Goldin (2012, p. 175).
[65] See Banting and Kymlicka (2002).
[66] Goldin *et al.* (2012, p. 175).
[67] Collier (2013, p. 63).

and complex research area full of controversies, the preceding account of those effects served to provide a basic notion not only of the processes at work, but also of the risks that are taken, and of the ethically relevant interests at stake when immigration occurs.[68] It became clear that immigration, especially in its rapid and large-scale form, comes with a series of serious problems for the host country. These problems must be taken seriously when examining the morality of immigration. The costs and risks on the side of the host country and its citizens must be weighed up carefully with the interests and rights on the side of the immigrants.

---

[68] Note that beside the here discussed economic and social effects of immigration, there are also environmental effects of immigration. Indeed, it comes close to a commonsensical assertion that *ceteris paribus* any increase in the number of people living on a particular confined territory will lead to that territory's resources being used (up) more extensively. So if a country of a population of 100 million has this population increased by 10 million further people (for example through immigration), then there will be 10 million further people who need space to live, who need resources like water, who consume, and who produce waste. One might object that such an increase could be offset if the whole of the population used available resources more sustainably, if more efficient technologies were available or if people were simply educated to consume less. But one limitation of such a counter-argument is obviously that it depends on optimistic assumptions: namely the hope that better technologies will be available, or that there will be more environmentally conscious people, etc. One author goes even further and argues that the environmental problems related with immigration are not limited to the country of destination. Rather, immigration into rich industrialized countries could also come with negative effects for the *global* environment and the climate (Cafaro, 2015, pp. 157-76). In Cafaro's view, this is so because in so far as immigrants tend to come from poorer countries in which people tend to emit far less, they are likely to adopt their new (and more prosperous) country's lifestyle and thus emit more greenhouse gases. So even though the global population may not change through immigration, what could change is the overall amount of global emissions and thus the rate at which further carbon is added to the atmosphere. For the case of limiting immigration based on environmental concerns, see Cafaro (2015); and for the opposing view that it is inappropriate to adduce environmental reasons when arguing for more restrictive immigration policies, see Neumayer (2006).

Part I

Normative Foundations: Basic Rights and the Ethics of Immigration

Two

Basic Human Rights and Correlative Duties

All human beings have *equal rights* to basic freedom and well-being, which are the necessary conditions for leading a minimally decent life. This principle of the equality of rights, which remains to be substantiated in this section, sets the fundamental normative standard for all moral debate in this work. The crucial questions with a view to the issues of immigration and exclusion are the following: how does the (potential) immigration of (a particular group of) outsiders bear on the equal rights of all parties affected by that migratory process? And how could one justify exclusion without denying the equal rights of the affected people? For exclusion seems to express the preferential (moral) treatment of a group of insiders. Or rather: to the extent that it is the insiders who decide to exclude outsiders, they seem to deny outsiders' moral equality. If this observation were true, exclusion would appear to be morally untenable. It would be inconsistent with the assumption that all human beings are from a moral standpoint fundamentally equal. So to argue for the general permissibility of a state's right to exclude, it will have to be shown that exclusion is in principle consistent with this moral equality. This is a central aim of this work's first chapters. In a nutshell then, the normative rationale to be adhered to for the present reasoning on the morality of migration and immigration is this: any differential treatment of morally equal persons requires — as a justification — an account of how this differential treatment is in principle consistent with the excluded persons' basic moral equality. What would constitute a fair treatment of climate-induced migrants will be assessed by drawing continuously on this and other first principles to be outlined in this chapter.

With these general remarks setting the agenda for the following sub-chapters and sections, I will now take a step back and outline in a more systematic way what it means to have (equal) basic rights and, just as important, what kinds of obligations they entail. Now, the normative starting point of moral reasoning in this work is a rights-based moral theory.[1] So what I intend to provide is a general characterization of such a rights-based moral theory, i.e. an outline of the main elements and conceptions needed to make sense of moral rights and duties. While this outline remains largely assertive, at least a few words on how to justify rights will nonetheless be added below. By then, I will have established the normative equipment with which to approach the question of what we owe to climate-induced migrants. This equipment will be further refined and readjusted in the course of applying it to questions of migration and climate change, and it will be enriched with a range of empirical assumptions needed to deal with such practical issues.

A rights-based theory as I shall set it as the normative standard here is built around the central assumption that there are certain *basic human rights*. Let us begin with an analysis of the central components of a conception of human rights. First of all, human rights are rights which persons have simply in virtue of their being human. All humans have them, which makes them *universal human rights*. And as we all have these rights equally, they are equal human rights. On the basis of the assumptions postulated so far, one may now distinguish universal human rights from "special rights". Both universal rights and special rights will play a role in this work. While special rights can be claimed "because of an act, event, or relationship of which a causal or historical account can be given",[2] universal human rights hold irrespective of such contingencies. They are firmly fixed in the common humanity and the dignity we all share—features which no human can lose or dispose of. Human rights are in this sense not only equal and universal but also *inalienable*.

This leads to a second characteristic of human rights, namely that they are basic in the sense that they designate a *moral minimum*. As such, they take priority over other rights in the moral universe which are not as fundamentally important as basic human rights. In cases where rights conflict, the more important human right to life will, for example, take situational priority over property rights. The point here is not only that the right to property is less important than the right to life, but that there is a strict hierarchy between them which disallows it to have human rights traded off against other moral rights which are

---

[1] As this is a work in *applied* ethics, I will not spend much time on discussing how the assumption that there *are* rights could be justified. This assumption has been made plausible by other authors, for example by Gewirth (1978), Shue (1996), and Steigleder (1992). This allows me to largely restrict myself in this work to an application of rights-based theory.

[2] Shue (1988, p. 688).

not basic. As Simon Caney sums it up, "human rights specify minimum moral thresholds to which all individuals are entitled, simply by virtue of their humanity, and which override all other moral values"[3] and even a communitarian theorist like David Miller agrees with this understanding of human rights when he notes that "the purpose of human rights is to identify a threshold that must not be crossed rather than to describe a social ideal."[4] But what is it about the substance or object of a right that makes it literally *basic*? According to Henry Shue's conception of basic rights, a right is basic when "any attempt to enjoy any other right by sacrificing the basic right would be quite literally self-defeating, cutting the ground from beneath oneself".[5] Similarly, Gewirth departs from the observation that certain goods are basic in so far as we need them for our very capacity to act at all. Freedom and certain basic physical dispositions are *necessary* goods, whose necessity for action no reasonable person can deny without contradicting herself. With a view to such basic goods, one may not only distinguish a procedural aspect (freedom) from a substantive (well-being) one. It also bears noting that both aspects can be understood either in an *occurrent* sense that focuses on the good's necessity for particular actions, and a more general, *dispositional* sense where the focus is on the good's necessity for controlling one's actions over the long run. This is how Gewirth exemplifies this distinction with a view to the procedural aspect of basic freedom:

> The loss of dispositional or long-range freedom, such as by imprisonment or enslavement, makes all or most purposive action impossible, while to lose some occurrent or particular freedom debars one from some particular action but not from all actions. Nevertheless, the loss of freedom in a particular case deprives one of the possibility of action in that case.[6]

This expresses a clear hierarchy. The dispositional availability of freedom and certain material goods are in principle more essential than the capacity to act freely in this or that concrete situation.[7] Indeed, this distinction between the occurrent and the dispositional (long-range) aspect of basic freedom and well-being as necessary goods for action will turn out to be absolutely crucial for later conceptual work on

---

[3] Caney (2010, p. 165).
[4] Miller (2016, p. 33).
[5] Shue (1996, p. 19).
[6] Gewirth (1978, p. 52).
[7] See Steigleder (1992, p. 145).

categories of migrants and duties toward them.[8] It is a distinction that allows for important normative insights when it comes to the question of what we owe to vulnerable people and refugees.

As a supplement to the human rights conception posited so far, I shall now add a few considerations on the question of *why* anyone should actually act in a morally required way and respect the human rights as just outlined. It is now that I will shortly touch on the question of justification, which Alan Gewirth calls the authoritative question of moral philosophy.[9] This question can be approached by departing from the crucial observation that every person must regard certain goods as necessary for her very ability to act, goods in which she will invariably have an essential interest, goods which she must want, and goods of which she knows that others must want in the same way. These necessary goods are procedural things such as security and freedom, and certain basic material (or substantive) things which she needs for her subsistence or basic well-being. Now if she thinks that she can lay claim to any rights at all, which Gewirth argues she justifiably and invariably will think, she will hold that there cannot be any things more urgent to which this right-claim may be directed than to the necessary conditions of the very possibility of her (successful) agency, which again she must want, and she will indeed have to acknowledge that all other people equally hold this right-claim then.[10] This reflexive approach of Gewirth to the authoritative question draws its plausibility from the fact that one cannot deny any of the argumentative steps without contradicting oneself. One is—for reasons of consistency—

---

[8] Note that by my understanding of this distinction, a loss of longer-range dispositional freedom does not necessarily imply the loss of occurrent freedom (a point where I appear to diverge slightly from Gewirth's characterization of that distinction in the above quote). What I take dispositional freedom to mean instead is that without longer-range freedom one cannot control one's behaviour and pursue one's longer-term life plans freely. To make clear that distinction and its application in this work, consider the following example: when you live on an island that will be submerged by the rising sea level only tomorrow, you may today still be free in the sense of occurrent freedom. But you are not free in the sense of dispositional freedom as you are—all else being equal—deprived of the possibility to conceive longer-range plans.

[9] The two other questions from which Gewirth distinguishes the authoritative one are the distributive and the substantive one. The distributive question asks whose interests the moral agent should consider favorably and further: "To which persons should the goods accruing from such consideration be distributed in actions and institutions?" (Gewirth, 1978, p. 3). The substantive question asks: "Of which interests should favorable account be taken? Which interests are good ones or constitute the most important goods?" (*ibid.*).

[10] See Gewirth (1978, p. 63).

required to accept each step of Gewirth's argument, so that denial comes at the cost of inconsistency, and consistency should matter a lot to reasonable people.[11]

For strategic reasons, I will not tie my own argumentation in this work too closely to either Gewirth's or any other particular moral philosopher's moral groundwork. I will henceforth just take for granted that there are certain basic human rights that every person can claim. Based on the above elaborations, this assumption can be made with some confidence by now. Moreover, it is an assumption that is, fortunately, confirmed by the reality of international human rights law, humanitarian law, and refugee law.

Much in line with this pragmatic strategy, some authors have proposed to skip the question of how to provide a final justification of human rights, which is what Henry Shue does in his book *Basic Rights* (1996). This is what Shue contends: if the moral subject assumes to have a right at least to *something* (which presumably she will, and which indeed no one would want to deny, for the negation of such assumption would mean she does not have any rights at all), then she must acknowledge that there are certain *other things* (i.e. the substances of basic rights) which are necessary for enjoying that *something* as a right, and she must then hold that she has a right to these *other things*.[12] This argument allows us to come back to an important feature of human rights which has already been mentioned above, namely their universality. The moral subject in the above argument will not only assume to have a right to something—an assumption whose denial is not only inconsistent according to Gewirth but which would indeed and fortunately be in plane contradiction with the mentioned realities of international law.[13] Rather, and here I draw on Gewirth once more, she must assume that all other moral subjects have such rights as well, because she must admit that all those other moral subjects are

---

[11] Apparently, the success of Gewirth's argument hinges entirely on this claim that one can indeed not deny it without contradicting oneself. For an elaborate and as I see it successful defence of Gewirth and this crucial claim, see Steigleder (1992).

[12] See Shue (1996).

[13] This actual codification of human rights suggests another justification of moral human rights, namely a practical one as for example Charles Beitz sets it forth: "A practical conception takes the doctrine and practice of human rights as we find them in international political life as the source materials for constructing a conception for human rights. [...] we take the functional role of human rights in international discourse as basis: it constrains our conception of human rights from the start" (Beitz, 2009, pp. 102-03). This passage of Beitz is similarly quoted in Weitner (2014, p. 39).

relevantly similar to her. They also have essential interests in certain goods necessary for their freedom, like security and subsistence, and they equally are dignified persons deserving of respect.

Finally, it needs to be asked how that understanding of a rights-based ethics can be applied to the reality of climate change and immigration. How can we meaningfully use it as an instrument to assess and evaluate concrete immigration policies? I shall leave the answer to this question of operationalization to Klaus Steigleder:

> Such a rights-based ethics focuses on the consequences of actions. The crucial question for us in evaluating the moral rightness of actions and of the circumstances of action is this: How do these actions and the circumstances of action impinge on the rights of all affected persons, do they infringe on these rights or not? Since the individual rights serve to protect goods that are to varying degrees important, or even indispensable, for the fundamental entitlement, and since every person has the same fundamental entitlement, in certain exceptional situations one person's more important rights may take priority over another's less important rights.[14]

Importantly, as Steigleder observes in the above quote, the focus of a rights-based approach as it is applied here is on the consequences of actions. Whatever policies and institutional settings or changes I come to propose, prior attention must have been paid to the way in which those policies affect the rights of people in actual practice. This focus on reasonably predictable consequences of policies and institutions resonates with the kind of *political approach* that was outlined in the introduction and that will be adopted in this work.

I will now (and particularly in the following sub-chapter) come to a concern with moral *duties*. Conceptually, for human rights to be effectively protected there must be duties that correlate with the assumed human rights. Concerning the structure of a right there is a person who holds the right, at least one person against whom this right is held, and a substance (or content, or object) to be protected through the right. We speak of claim-rights then in the sense that they can be pressed and demanded against other persons, and indeed the right-holder can insist on being granted the substance of the right, for it is her due, and other persons owe it to her. Following Henry Shue's account in his book *Basic Rights* (1996), I will here assume rights to correlate with at least three types, or levels, of duties.[15] There will be

---

[14] Steigleder (2012, p. 6).
[15] Note that there are substantial deviations from Shue's account in the way I will characterize (and label) the different types of duties. He makes no use of a model of moral division of labour between states (MDL) under the second type

some substantial variation from Shue though, especially in so far as my account will be centrally concerned with duties of *states* (at least on the second and third level). I will blend Shue's account with other morally-normative considerations in such a way that the resulting *three-fold scheme of duties* is adapted to fit the needs at hand—an application to the realities of climate change and migration. To provide some orientation in advance, this is the scheme to be developed in this chapter:

I. The general duty to avoid violating human rights.
II. The assigned (special) duty (of states) to protect human rights.
III. The duty (of states) to assist people deprived of human rights:
   1. The duty (of states) to help the unprotected.
   2. The duty (of states) to assist the victims of one's own harmful actions.

What follows now is a systematic characterization and discussion of each of these duties. I start with the first-level duty that correlates with human rights.

2.1. The General Duty (I) to Avoid Violating Human Rights

Just like rights, duties can be either negative or positive. While negative duties require the duty-bearer not to perform a certain action toward the holder of a correlative negative right, positive duties require the duty-bearer to perform a certain action toward the holder of a correlative positive right, at least under certain circumstances. Let me now make a few further remarks on negative duties. Note first that negative duties are invariably universal, meaning that everyone bears them. This is what Shue has to say on the universality of negative duties:

> Negative duties—duties not to deprive people of what they have rights to—are, and must be, universal. A right could not be guaranteed unless the negative duties corresponding to it were universal, because anyone who lacked even the negative duty not to deprive someone of what she has rights to would, accordingly, be free to deprive the supposed right-bearer.[16]

Note further that negative human rights-related duties are general duties, so that we owe them to other humans simply because they are

---

of duty (II), nor does he introduce explicitly a conception of corrective justice (CJ) under the third-level duty (III-2).

[16] Shue (1988, p. 690).

humans and we thusly owe them to all those other humans.[17] We owe certain negative duties to every one person in the world. Fortunately, it is indeed (at least in theory) not difficult to fulfil one's negative duties toward an infinitely large group of people. As Shue observes, "I can easily leave alone at least five billion people, and as many more as you like."[18] So with negative rights being rights held against every other person, any such other person is left with a duty not to deprive the right-bearer of her negative right not to be harmed. Needless to say, the negative duty to avoid depriving others of the substances of their human rights applies not only to individuals but also to all other accountable entities—such as companies, countries, or states—which through their actions potentially affect other people's basic rights to freedom (or security) and well-being (or subsistence).[19]

I will now come to the second-level duty one should accept if one takes it seriously that other people have basic rights and if one hence wants them to actually *enjoy* the substances of such rights. Duties must be *assigned* to accountable agents in order to make sure that the rights which ground such duties are *effectively secured*.

2.2. The Assigned Duty (II) of States to Protect Human Rights and the Right to Self-Determination

It would be illusory to assume that (universal) positive rights[20] can ground general positive duties: it would be absurd to hold that the bearer of a certain positive human right has a right to the object of that positive right against every other person in the world. Universal positive rights cannot entail general (positive) duties in the same way that universal negative rights entail general negative duties. The positive duties that correlate with positive rights are always conditional on the abilities and resources of potential duty bearers to actually fulfil the positive duty. Shue's sober observation that "the total amount one can owe to others must be limited simply because one's total resources are limited"[21] is then just a different way of saying that one cannot be obligated beyond that which one is capable of doing—a general truth encapsulated both in the Latin phrase "ultra posse nemo obligatur" and

---

[17] By Shue's terminology, a *duty* is general when "it is owed on some ground independent of specific acts, events, and relationships" (Shue, 1988, p. 688), with the alternative being a special duty (*ibid.*).
[18] Shue (1988, p. 690).
[19] See Weitner (2013, pp. 26–27).
[20] For example, the positive right to basic subsistence that everyone has.
[21] Shue (1988, p. 690).

in its English variant of "ought implies can".[22] The core normative point here is that one has a duty to help others only as long as one can do so at "no comparable cost to [one]self".[23] It may be hard to specify what "at no comparable cost" means, and specifying this for the case of climate migrants is a major task of this work. Often, we will indeed have to draw on a dynamic risk-ethical perspective in order to solve this difficult issue of specification. As a first approximation, however, a general reason why the costs on my side as the duty-bearer invariably have the potential of becoming too high is that, with rights being equal rights, I also have (basic) rights which entitle me to "consume at least some of my resources for myself".[24] How many of those resources I am entitled to consume for myself may then depend on concrete contexts, for example on whether a prior wrongful act is involved, which might shift—to my disadvantage—the line of what may be reasonably demanded of me.

In practical terms then, the problem is that even though all of the global poor have a positive right to subsistence, I clearly do not have a duty to help each of them, simply because I do not have the resources needed for that purpose. What can be observed is thus that universal (positive) rights practically cannot entail general duties. Rather, what they entail in Shue's words is the ideal of *full coverage*: "Full coverage can be provided by a *division of labor* among duty-bearers. All the negative duties fall upon everyone, but the positive duties need to be divided up and assigned among bearers in some reasonable way."[25] The key term to be emphasized in the above quote is "division of [moral] labor". The aim is that for every person who holds a positive human right, accountable agents must be found who are assigned the task of fulfilling that right. In other words, some specific agent or group of agents—agents who are capable of effectively fulfilling that right— should get assigned the duty that corresponds to the right in question.[26]

For reasons that remain to be spelled out in the next section, I will assume that in the current world divided into states it is individual states that get assigned and then bear the primary responsibility of protecting on their own territory the rights of their citizens.[27] More

---

[22] See Griffin (2008, pp. 72-73).
[23] Gewirth (1978, p. 217).
[24] Shue (1988, p. 690).
[25] *Ibid.*
[26] See Shue (1988, p. 689).
[27] For a first pragmatic justification of this view, see Nickel (1983, pp. 36-41). According to Nickel, who at the moment sees "no genuine alternative" (Nickel, 1983, p. 35) to the present system of sovereign states, the decisive pragmatic

precisely, the citizens within an individual state will have *special duties* toward each other, so that in actual fact it is not so much the state that has obligations toward its citizens but rather the citizens—together constituting the state—that have obligations toward one another, with the state coordinating the effective realization of that mutuality and reciprocity.

It turns out then that the here concerned second-level duties (II) of *protecting* others' rights are realized (at least mainly and standardly) as special duties among citizens. What I theorized as duties *assigned* to states standardly take shape then and are effectuated as *special* duties within those states. Finally then there evolves an overall system of moral division of labour in which all rights are effectively covered. The result, ideally, is full coverage.

2.2.1. From the Protection of Human Rights toward the Right to Self-Determination

In this section, I will make plausible the already suggested claim that it is among individual states that moral labour should be divided because, as I will contend, states are the units where such labour can be divided relatively (or presumably even *most*) effectively. Within each state, a community of citizens—on the basis of mutuality and solidarity—provides for the effectuation of each other's rights. This is what Alan Gewirth has to say on the effective mediation of duties through and within individual states:

> [I]t is the community that protects and fulfills these rights, especially for the persons who do not effectively have them, on the basis of the mutuality that characterizes human rights and the consequent solidarity of the society whose members are brought together by their recognition and fulfillment of common needs. Thus rights require community for their effectuation, and community requires rights as the basis of its justified operations and enactments.[28]

---

reasons for assuming this responsibility of states (i.e. their governments) to protect their citizens' rights are twofold: if a government were to relegate that responsibility to another state or international organization then it would be doubtful to what extent that government would actually remain a self-determined "government": "Protecting people's rights is such an essential role of governments that a government deprived of that responsibility for upholding the rights of its people would be no government at all" (*ibid.*, p. 36) The other reason is that the government on a given territory is the most effective agent to successfully fulfil this task of effectuating the citizens' rights. It is closest to the right-bearers and has the necessary means for effectuating their rights at its disposal.

[28] Gewirth (1996, p. 97).

Indeed, to say that "rights require community for their effectuation" remains somewhat underdetermined. Rather (and this is of course what Gewirth means), what the effectuation of rights requires is in the concrete case a *particular* community. Any conception of justice and human rights, universal though it (rightly) pretends to be, is invariably realized within particular communities of people who together constitute what Beat Sitter-Liver calls a *room of justice*.[29] Such rooms of justice are constituted and maintained by citizens according to their conception of justice. As a concrete polis, and on the basis of mutuality and solidarity, they push forward their conception of justice within that room. It stands to reason that the citizens within that room allocate rights and obligations first and foremost among themselves; they have special duties toward each other. To deny that they have such special duties toward each other means to call into question the very idea of particular communities as realizing their own rooms of justice within which to effectuate rights. But as long as there is no globally shared conception of justice that would allow us to tear down the walls of such particular rooms of justice and replace them with a more universalistic scheme, any denial of special duties within the concrete polis is tantamount — at least practically — to venturing the only effective system of rights protection we currently have.

The insight of the relative effectiveness of individual states (and their respective rooms of justice) in securing the human rights of (their) people suggests — globally speaking — that each state within the international system of states be assigned the *special responsibility* to secure the human rights of its own citizens. Within the respective room of justice, each state has to fulfil this task by establishing effective economic, political, and social institutions, which are possible only in and by means of a particular community. It is through these institutions that human rights are to be, as Shue puts it, *socially guaranteed* against standard threats, which "means, in effect, that the relevant other people have a duty to create, if they do not exist, or if they do, to preserve effective institutions for the enjoyment of what people have rights to enjoy".[30] While Shue only holds that for a basic (human) right to be effectively secured it needs to be socially guaranteed, my assumption here goes clearly beyond Shue's account. It concretizes Shue's account in the important respect that *individual states*, as parts of a larger system of states that divide moral labour among them, are specified as the agents that are each assigned the task of protecting the

---

[29] See Sitter-Liver (2003, p. 40).
[30] Shue (1996, p. 17).

human rights of their respective citizens. This is clearly a normative statement rather than a description of the way in which states as parts of the international order conceive of themselves.

Indeed, the model of moral division of labour between states (MDL) is here understood as an ideal that makes normative sense of the reality of the state system. Individual states and the respective political communities of which they are made up provide for the coordinative arenas in which the rights of people, who in those communities figure as citizens and members, are effectuated. A world covered by such states would result in the previously suggested ideal of full coverage (of human rights). The realization of this ideal should remain the aim of the states that together make up for the system of states, and it begins with each individual state fulfilling first and foremost its *own* share of human rights protection. This ideal is the best hope we have today. Under realistic assumptions, there must be state institutions of a certain kind in order to protect rights, for there is currently no feasible effective alternative. This lack of alternatives only serves to underpin the notion that, as Gewirth puts it, "human rights include one's right to be a member of political community and to be helped thereby to develop one's full humanity as an agent."[31]

Now, how do the previous elaborations on the necessity of states relate to the question of whether or not legitimate[32] states—and more precisely their citizens—should have a right to self-determination? Arguably, in order to fulfil the primary task of protecting their own citizens' human rights, it is of fundamental importance that states have the right to protect the state's institutions and control the inner work-

---

[31] Gewirth (1996, p. 69). This consideration will turn out to be crucial for later discussions of what is owed to refugees. It will remind us that refugees have a right to be members of some state in order to develop their *full humanity* as free agents. But the consideration is ambivalent: it will be worth recalling that existing citizens have a human right to their own functioning state institutions as well, which through the admission of very large groups of refugees could be destabilized.

[32] Note that the legitimacy of an individual state is here taken to depend on its efforts in protecting its own citizens' human rights (which include a right to a minimum of democratic self-determination) and in not violating the rights of other people (see Buchanan, 2004, p. 187). In fact, all the individual state has to do is to fulfil its share of human rights protection and not hinder other states from fulfilling their share, and as Miller rightly notes, "[a] state that fails to do this, despite having the necessary resources, is to that extent an illegitimate state" (Miller, 2016, p. 34). So by the here proposed understanding, the legitimacy of an individual state does not depend (also) on the legitimacy of the state system, as for example Owen (2016) argues.

ings of society that allow for the protection of human rights in the first place. The right to self-determination guarantees that such human rights-related efforts and successes are not undermined or impeded through the meddling of third parties.[33] Indeed, the justification of a legitimate state's entitlement to self-determination comes from two mutually supportive directions: a legitimate state should have a right to self-determination *so that* it can protect the community and the conditions through which it can effectuate the human rights of its citizens, and it should have a right to self-determination precisely *because* and to the extent that it does fulfil this task of protecting its citizens' human rights.[34]

2.2.2. On the Justification of States and the State System

This section has the important function of further substantiating the general view that states are justified and that they should figure as the units among which to divide moral labour (and which, thusly, are standardly assigned the duty II to protect). The focus of this justification is laid on aspects other than those already discussed in the previous section. What will be provided is a deeper understanding of the role of, and the need for, (legitimate) individual states in realizing basic human rights. It will be against the backdrop of this understanding that my later discussion of the morality of immigration control will take place.

Now, (democratic) states ideally fulfil the function of distributing—according to fair and agreed upon criteria—the benefits that are produced within the society of the respective state, and of distributing

---

[33] See Donnelly (2013, p. 157).
[34] However, the right to self-determination appears to come as a double-edged sword. It not only implies that legitimate states and their citizens have a right that third parties not meddle with domestic affairs. It also implies that the citizens of the state are, at least to some extent, responsible for the consequences that follow from the self-determined decisions they make (i.e. the principle of "self-responsibility"). So what should be noted here is the possibility of a *tension* between the two normative principles of human rights on the one hand and self-determination, or more specifically self-responsibility, on the other hand. If two parties are in dire need and require my assistance, and if in this particular situation I have the means to help only one of them so that I must decide whom to help, it might well be a relevant information (beside other possibly relevant considerations) that one party but not the other brought himself into such dire straits through his own (unwise) choices as a self-determined and self-responsible agent. The same will hold for self-determined collectives, all else being equal. Evidently, this insight has potential implications for immigration policy.

according to fair and agreed upon criteria the duty to contribute to the state's institutions through taxes. This duty would otherwise be left to voluntary groups, and this in turn would open the doors to free riders.[35] So beside the main justification that states protect human rights, a further justification of states is their *coordinative function*: as members of communities and their respective states, people "establish property and taxation regimes which are important for improving their economic well-being, both individually and collectively, and, especially, they develop processes and institutions to solve collective problems".[36]

Such general justifications of states notwithstanding, one could of course object that other ways of organizing the world to the effect that (1) human rights would be effectively secured and (2) coordination problems would be solved are imaginable. Why not go for a world state? Or why not abandon states altogether and organize social life at much smaller units without the potentially destructive power of states, in an anarchical way for example? In this section I provide two additional and indeed rather *pragmatic* justifications for focusing on states and for accepting the state system as a fact that one should not question in too fundamental a way. (Note that in the following chapter 3 on the morality of immigration, I will then explore a further *substantive* justification for the assumption that *nation*-states are appropriate and effective units to take on special responsibilities, a justification that takes as its point of departure an acknowledgment of the importance of culture and shared values as the functional substrate of community.)

Regarding the first pragmatic justification, I will draw on a rationale that John Rawls offers in *The Law of Peoples* (1999). In a famous passage, he provides a powerful reason for the here proposed view that it is appropriate to hold individual units like states to be the bearers of special responsibility. His focus is on the way in which a system of individually accountable states sets collectively advantageous incentives:

> An important role of a people's government, however arbitrary a society's boundaries may appear from a historical point of view, is to be the representative and effective agent of a people as they take responsibility for their territory and its environmental integrity, as well as for the size of their population. As I see it the point of the institution of property is that, unless a definite agent is given responsibility for maintaining an asset and bears the loss for not doing so, that asset tends

---

[35] See Gewirth (1996, p. 59).
[36] Moore (2015, p. 209).

to deteriorate. In this case the asset is the people's territory and its capacity to support them in perpetuity; and the agent is the people themselves as politically organized.[37]

Rawls continues by arguing that the citizens forming the state "are to recognize that they cannot make up for their irresponsibility in caring for their land and its natural resources by conquest in war or by migrating into other people's territory without their consent".[38] This verdict of Rawls resonates with the above assumptions on self-determination and the emphasis of self-responsibility that it comprises. In fact, he complements such previous assumptions by putting the focus on *incentives*. More to the point, he stresses that the insistence on the importance of self-responsibility, and therewith on the importance of bearing at least to some extent the consequences of one's own actions, sets exactly those incentives that are most welcome from an overall human rights perspective. Rawls calls for the set-up of a system of states which provides for that incentive structure that is likely to make a system of states an efficient institution to protect human rights globally. This is, in his view, effectuated by making a particular state's territory the asset of that state's citizens.[39] Note that the alternative proposal would be to make the world's territory a global common, an alternative whose efficiency in securing basic rights globally is indeed most questionable. After all, Rawls' emphasis on incentives further supports not only the assumed normative ideal of MDL but, more generally, provides for a further reason — a pragmatic one, if you like — for accepting states. As parts of a system of states the individual states will face the right incentives for taking care of their resources, provided they know that they cannot externalize the negative consequences of their own unwise decisions. How and whether this assessment bears on the ethics of immigration policy and on the discussion of climate migration remains an open question to be discussed later.

I will now turn toward to a second pragmatic justification for focusing on states. As a crucial observation, the world *is* organized into individual states, so a model of *states* as dividing moral labour between them provides for a theoretical frame that at least partly mirrors real-world constellations. There are real people today who find in their state the only guarantor of their rights. German legal theorist and former

---

[37] Rawls (1999, pp. 38–39).
[38] Rawls (1999, p. 39).
[39] This view seems to imply the right to decide on immigration onto that territory (i.e. its asset), because only by controlling how many people use that asset and live on that asset can citizens exert control over whether or not that asset is overused or not, and by extension whether it is mismanaged or not.

member of the constitutional court Udo di Fabio rightly observes that in the absence of today's states the whole idea of rights would lack its "material substrate", and that there would be no agency that the individual right-bearer could go to in order to effectively claim his rights against others.[40] What would the alternative be today? Should we risk the transition to another arrangement and leave behind that order which provides for peace and the protection of human rights in large parts of the world? In the world as we find it today, the acceptance of states can—from a human rights perspective—hardly be seen as the bitter pill that needs to be swallowed. The alternative is not clear and, whatever it is, the transition to it would be risky, and if the called for alternative were an anarchical world without the strong monopolies of power that states constitute in order to effectuate human rights, then the transition toward that alternative would mark a safe way to ruin, rather than a risk.

The evident recognition that the real world confronts us with states and that at least some of them work reasonably well informs the *political approach* as pursued in this work. This approach urges us to work with the realities we face, to analyse them, and to make normative proposals that take seriously the constraints those realities present us with. Especially for dealing with questions of immigration it will prove suitable to take the reality of states for granted, for immigration is an inherent concern *only* in a world of states and borders. It is in this vein that David Miller proposes to swallow a "considerable dose of realism here" simply because "the immigration *issue* would either disappear altogether or at least become much less pressing in a world that was configured quite differently from our own."[41] So to disengage with the paradigm of states and the state system would impede the formulation of *practically useful* moral criticism right from the start. However, to adopt a normative framework compatible with the existing paradigm of states and borders does not imply any kind of undue conservatism. The political approach notwithstanding, the focus on states is no absolute, but remains open to critical re-examination.[42] More still, if its flaws (and thusly those of the state system) turn out to be too fundamental, i.e. too irreconcilable in practice with the principle of all people's equal moral rights, then it would indeed seem unavoidable to criticize this reality more fundamentally.

---

[40] Di Fabio (2005, p. 82).
[41] Miller (2016, p. 17).
[42] This approach is similarly used by Buchanan (2004, pp. 53–59).

After these complementary remarks on the justification, the necessity, and the responsibility of states, I take to be well-founded the view that individual states should be the units among which to divide moral labour (and which should standardly get assigned the duty II to protect in order to ideally achieve full coverage of human rights).

2.3. The Duty (III) of States to Assist People Deprived of Basic Rights

What remains to be outlined is the last of the three types of duties that correlate with basic human rights, namely the duty (of states) to assist people deprived of basic rights. This third-level duty becomes relevant only when, or to the extent that, the provisos of the first- and the second-level duties turn out to be ineffective or when they were disregarded by some duty-bound party under I and II.

Now, there is a general duty to help people deprived of the substances and freedoms of basic rights quite simply because they have by virtue of their humanity a right to these substances and freedoms. This duty is invariably a conditional one though, for it can only be fulfilled in so far as the helper has the means to help without comparable cost. In light of the above account of basic human rights, I consider it to be beyond doubt that such a duty to assist exists. That one is under a conditional obligation to help people in dire need indeed amounts to a default mode for all subsequent reasoning. The following account of why to help people deprived of human rights goes beyond that basic assessment though. In that account, it is *states* that have a duty to assist people from or in other states who find the substances of their basic rights continually and seriously at risk. If individuals have the duty to help other individuals in case of dire need, then by extension states—which are made up of such duty-bound individuals—also have a duty toward needy others, especially when it is states rather than individuals that have the means to provide effective assistance. Indeed, it is certainly only states or groups of states that have the means and that can act in sufficiently coordinated ways so as to deal with a humanitarian problem like the global refugee population.

This is how I will proceed: in the first section (2.3.1), the general duty to help third parties will be integrated into the previously assumed normative model of the system of states as a system of moral division of labor (MDL); in the second section (2.3.2), the concern is a different one. Here, I will look at those circumstances in which the potential helpers (i.e. states and their component individuals) played a relevant causal role in bringing the persons in need of help into that very situation. In such circumstances, the thrust of the distinct moral principle of corrective justice (CJ) *adds* to the normative principle of human rights.

2.3.1. The Duty (III-1) of States to Help the Losers of the State System

Arguably, the proposed MDL-model in which states are responsible for fulfilling *primarily* their own assigned share of the human rights-related duty to protect (II) has one important implication: when with a view to any particular dysfunctional state that assignment of the duty II to protect is not successful so that the basic rights of people in such dysfunctional countries remain widely unprotected, the concern with basic rights urges other states to find an alternative way of protecting their rights. From the perspective of the world's state system those people whose rights are not protected effectively by their own state can be conceived of as losers of the state system. A remedy must be found for those people who find their state (which after all they never chose) unable or unwilling to protect their rights. As Hannah Arendt puts it, any subject under the state system "can lose all so-called Rights of Man without losing his essential quality as man, his human dignity. Only the loss of a polity itself expels him from humanity".[43] States that acquiesce in the reality that some people are ultimately left with effectively *no rights* at all can hardly be seen to act in accordance with the principle that all people possess equal moral rights. The minimum effort that other states under the state system have to make in the face of such deprivations is a *credible commitment* to find solutions for such an untenable situation. This is a helpful political framing. But the fundamental moral point is that there is a group of people who are utterly helpless and unprotected, and they have a right to be assisted.

What is needed are provisions for those people who do not find their human rights protected due to an inability or unwillingness of their own states to fulfil the assigned share of moral labour. Where individuals, as members of the overall state system (which they invariably are), find their basic rights violated and permanently unsecured, other states would have to step in and help those people, provided those affected individuals cannot—as citizens of their own states—self-organize and help themselves. To be sure, to self-organize and do away with the refugee-producing circumstances themselves clearly remains the primary responsibility of the affected citizenry. Such emphasis on the primary responsibility of citizens, however, is largely neglected in current ethical discussions on refugees, where the latter figure mostly as helpless victims. There can be no doubt though

---

[43] Arendt (1968, p. 279).

that where assistance by third parties (through asylum) is indispensable, it must be provided by states in a position to help.

Complementary to this basic humanitarian necessity to help, one may contend that what urges states to make such commitment to assist distant strangers and refugees is that the world's countries and their respective people are connected via complex international relationships in the economic, political, and cultural realm. As Charles Beitz frames it, they all participate in an international scheme of cooperation.[44] The recognition of this relatedness and indeed interdependence appears to have *some* moral relevance simply in so far as the reality of cooperation is also of relevance *within* countries, where (stronger) relatedness and mutuality is often considered to ground special duties (II) among citizens.[45] So one could argue that in cases where protection does not work at that inner-state level, affected people should be assisted by other states (and their people) because those affected individuals are people with whom the helping countries are connected in a global scheme of cooperation and from which those helping countries generally benefit. To some, this may be an appealing view in support of the duty to assist refugees. In the present account, it amounts only to a complementary view though. The bedrock view remains that states ought to assist the losers of the state system simply because and to the extent that they can do so at no comparable cost. They ought to help people who on an ongoing basis do not enjoy the substances of their basic human rights (III-1).

2.3.2. Harm and Compensation: The Duty (III-2) of States to Assist the Victims of One's Own Harmful Actions

An understanding of corrective justice, and first and foremost of what actions or what kinds of conduct justify *duties of compensation*, is central to this work's analysis. Working with only a few rather neat distinctions, I shall at this stage try to keep as simple as possible the question of when compensation (i.e. any action or measure that aims at restitution or repair) is someone's due. Later, once it comes to applying these corrective justice-related distinctions, I will refine them on a case-by-case level. On a most abstract and most general level, person A must

---

[44] Beitz (1979, pp. 143–53).
[45] "If social cooperation is the foundation of distributive justice, then one might think that international economic interdependence lends support to a principle of global distributive justice similar to that which applies within domestic society" (Beitz, 1979, p. 144).

provide compensation to person B when person A's conduct was "contrary to B's rights".[46]

Borrowing from Jules L. Coleman, I shall distinguish conceptually between two ways in which someone's action or conduct can be *contrary to another party's rights*, or, in other words, in which such action or conduct "*invades*" that other party's rights. The first is a *violation* of a right, which denotes a "wrongful invasion of a right". The second way is an *infringement* of a right: here the invasion of the right, i.e. the acting contrary to the right, is justifiable and thusly not wrongful in the sense of a violation.[47]

Now, the crucial assumption for this work is that in principle both forms of conduct contrary to a right — violation and infringement — can ground a valid claim of compensation on the side of the person affected by that conduct. So, to use Coleman's example, if A, out of necessity, appropriates B's insulin, then A's conduct may be justifiable, but it nonetheless constitutes an infringement. It is "contrary to the demands" that B's rights impose on A, so that the very fact of B's rights having been invaded is sufficient to ground a valid claim on the side of B to some kind of repair from A.[48]

Moreover, for compensation to be due I shall assume that it is not necessary that some party — in consequence of some rights invasion — has suffered an actual and tangible loss (or setback to the substances of basic rights). Rather, and here I draw once more on Coleman's conceptual and analytical clarity, the assumption is that compensation can also be required when someone's conduct, rather than actually and evidently causing a loss, is not adequately *respectful* of another party's rights. This lack of respect also counts as a behaviour that is contrary to rights. Importantly, such disrespectful conduct can also include someone's imposition of a relevantly severe *risk* to another person. Now, if such disrespectful conduct (like a relevantly severe risk-imposition) took place, then — in order for such "conduct to count as respectful of the victim's rights"[49] — compensation is required. This compensation could, for example, consist in anticipating the possibility of harm and in reducing the potential victims' vulnerability to the harm once it occurs. Where it is taken, such anticipatory action (which I will later in this work frame as ex-ante compensation) comes with the effect that the conduct of the risk-imposing party counts as at least *more* "respectful of

---

[46] For this general assessment, see Coleman (1992, p. 300).
[47] The distinction between violation of a right and infringement of a right is similarly used by Thomson (1986, pp. 49–65).
[48] See Coleman (1992, pp. 299–302).
[49] Coleman (1992, p. 284).

the [potential] victim's rights". So where such anticipatory measures were taken (or: where such "ex-ante compensation" was provided for), the forcefulness of *later* claims for (*ex-post*) compensation, i.e. in case the risk materializes, is at least reduced in degree, because the earlier anticipatory (or ex-ante compensatory) measures offset the disrespectfulness of the risk-imposition in question (and it is that disrespectfulness which would have grounded the duty to compensate ex-post).[50]

These introductory considerations of what it means to act against others' rights lead to the introduction of the moral principle of corrective justice (CJ). If an agent imposes on another person a human rights-related harm (or if through a certain conduct she relevantly contributed to such harm or increased the risk of such harm), thusly acting against the rights of the bearer of human rights, then under corrective justice a powerful duty is imposed on that agent to compensate for the harm (or more precisely for the loss, which is the object of the harm).[51] To be sure, there is no need to hold that the duty to compensate "follows as a matter of logic from the nature of what it means to have a right".[52] Rather, the principle of corrective justice is assumed to be instrumentally justified here. It is essential for effectively securing the substances of basic rights in practice. The invasion of one's rights must invariably be answered by a duty to compensate and, where possible, to make good for that invasion. This distinct normative principle of corrective justice is then justified by the overall concern with effectively equipping people, against all odds, with the substances they have a right to.

The duty to compensate (III-2) is arguably more powerful than the duty to help (III-1): while it is already a moral bad not to help some unrelated stranger although one is in a position to do so at no comparable cost, it is clearly more of a moral bad not to help someone although, and in addition to the fact that, one has oneself caused this harm to that someone: "Almost tautologically, then, the person who omits to aid another has failed less badly than the person who has not rectified a harm he has created, because the latter has both failed to rectify the damage he has done and failed to aid."[53]

---

[50] The idea is that in order to prevent over-compensation or "double-counting", any claim to *ex-post* compensation may be reduced by that which was provided already as *ex-ante* compensation. For a thorough discussion of the problem of double-counting, see Finkelstein (2003).
[51] See Adler (2007, p. 1858).
[52] Coleman (1992, p. 317).
[53] Lichtenberg (2014, p. 26).

Moreover and in addition to this relative normative strength, the duty to compensate is also characterized by a remarkable stringency which arises from the fact that it often possesses the property of being both a special and a perfect duty: it is *special* in the sense that it is "owed because of an act, event, or relationship of which a causal or historical account can be given"[54] — which tends to make the duty-bearers more naturally identifiable than is the case with other types of duties we have come across (especially III-1) — and they are *perfect*, meaning that at least in theory they are owed to specifiable, harmed individuals who can claim a right to the object of compensation.[55] Thus, compared to the assigned special duties under MDL, the duty to compensate is, though also a special one, assigned in a much more natural way. It requires that where A invades B's rights, A cannot get away with this disrespectful conduct.

The above account of duties that correlate with basic human rights finally results in the following three-level structure of duties.

I. The general duty to avoid violating human rights.
II. The assigned (special) duty of states to protect human rights.
III. The duty of states to assist people deprived of human rights:
   1. The duty of states to help the losers of the state system.
   2. The duty of states to assist the victims of one's own harmful actions.

This structure of duties and the underlying conception of basic human rights henceforth provides for the substantive framework on which I will draw to deal with immigration policy generally and climate migration specifically. However, in order to operationalize this framework successfully in a world in which we often do not know how our actions and policy choices bear on the rights of all involved and at least potentially affected parties, the rights-based approach as just established needs to be enriched by a risk-ethical perspective. This complementary perspective will render the approach sensitive to the uncertainties that pervade our world.

2.4. Discussing Rights and Duties from a Risk-Ethical Perspective

As already observed, the focus of a rights-based ethics must invariably be also on the *consequences* of actions for the rights of others. When reasoning on policies and institutional settings, attention must be

---

[54] Shue (1988, p. 688).
[55] See *ibid*.

directed to the way in which the rights of people will be affected by those policies and institutions. It was mentioned before that the approach that is taken in this work's morally-normative analysis of immigration policies is a "political" one or, if you like, a realistic one. What this means in essence is that I take very seriously how the proposed ways of dealing with immigration in general, and with certain groups of migrants like climate migrants in particular, bear on the rights of all involved parties. A proposed immigration policy that, when implemented by a particular state, leads to the breakdown of other but apparently not unrelated institutions of that state—like the labour market, the system of social welfare, or indeed the functioning of the market economy—can hardly be called a practically useful proposal. Does the state that acted upon that policy proposal have to admit in the aftermath that it was *wrong* to implement that policy?

Whether one can say of a given policy that it was the right measure to take or not depends on the informational basis the decision maker had *before* taking that measure rather than *after* taking it. In technical terms, the assessment of a decision should draw on the available information *ex-ante* and not *ex-post*, so people who are taking the relevant decision are pressed hard to look carefully at the available information on the decision's possible effects. This makes it possible to hold the decision makers responsible when the decision turns out to have undesirable consequences.

The problem of the decision makers is generally an epistemic one: they have no certainty about the effects of their decisions. The citizens and their elected decision makers simply do not know with certainty what results a given immigration-related policy or institutional set-up will yield. What they have to work with are indications and clues, the best available evidence on how the policy interacts with people, processes, and institutions in the real world. What they have to work with are *risks*, i.e. the realistic possibility that a negative result becomes reality. In Steigleder's words, risks are "possible effects or events that we as agents have to assume due to our epistemic situation".[56] Arguably, the decisions on the design of immigration policies are risky: they can have unintended and most undesirable effects for some people. So when the approach taken in this work is a *political one* that judges the goodness of a given institutional proposal by its realistic (i.e. reasonably predictable) consequences, then one cannot but assume here a *risk-ethical perspective* that, whenever needed, informs and complements this work's morally-normative examination. In the following, the indispen-

---

[56] Steigleder (2016, p. 261).

sable tool of a rights-based risk ethics will be outlined and it shall be indicated how such a risk-ethical perspective can be made applicable to the relevant issue of (climate) migration.

The risk-ethical perspective as I will introduce it now largely draws on Steigleder's conception of a rights-based ethics.[57] So first of all, I will settle the question of how a rights-based approach can deal with risks[58] in the first place. If one takes the equal human rights of all involved parties seriously, then how can one approve of the risk that a certain action, i.e. in our case the decision to implement a certain immigration policy, harms some of those involved parties and thus invades their rights? As Steigleder observes in this vein, rights-based theories tend to overreact when confronted with risks in so far as they "have a sort of natural tendency to prohibit all risk impositions. For if one has a right not to be harmed in certain ways, one also seems to have a right of not being exposed to the risk of being harmed in such ways."[59] The general solution of this problem lies in the recognition of the *equal rights* of the *agents* of the risk imposition and the *recipients* of the risk imposition. In other words, the people who impose a risk on others through a certain decision or policy have the same rights as those others who are potentially affected by that policy. While permitting all risks would mean to pay insufficient respect to the recipients' rights, a prohibition of implementing any policies that may mean risks to others would not only have absurd consequences but would indeed mean to take insufficiently seriously the rights of the agents:

> A rights-based ethics must pursue a dual perspective of both risk toleration, which focuses on the freedom, the plans, and the chances of agents, and risk elimination, which focuses on the danger to the objects of the rights of the recipients of risky actions. The focus on the equality of rights of both agents and recipients is the key for finding out which risk impositions must be tolerated and which risk impositions must be prohibited.[60]

---

[57] The basis of my account will be Steigleder's conception of a rights-based risk ethics as he unfolds it in a recent article (2016). In this article, the reader finds more detailed elaborations on how a rights-based ethics can deal with risks, as well as further relevant conceptual distinctions and applications.

[58] In following Steigleder, I will here "not make use of the established distinction between risks and uncertainties but will take risks as a generic term covering both risks in the strict or technical sense and uncertainties" (Steigleder, 2016, p. 261).

[59] Steigleder (2016, p. 261).

[60] Steigleder (2016, p. 262).

In a warming world, governments will have to prepare for the case that the number of migrants and refugees will increase. This reality will then urge them to make decisions that will impose risks on people, risks which depending on the concrete shape of such decisions will be distributed among those people in different ways. It remains an open question at this stage which immigration-related decisions could turn out to be risk-ethically relevant. It will turn out that there are, often, no win-win situations in the context of migration—a reality that many contributors to the current normative debate fail to acknowledge. People will react to immigration in certain ways that moral philosophers may discard as morally wrong. But their real attitudes must (at least partly) inform the risk-ethical perspective as just outlined.

After all, the basic theoretical instruments have been established now: the authority of basic rights and their corresponding duties, the plausibility of granting political communities (in which these rights are effectuated) the right to self-determination and of holding them to be self-responsible, the thrust of corrective justice, and finally an understanding of a risk-ethical perspective which reflects the awareness that we never have certainty on the ways in which our choices influence the very rights we set out to protect.

Three

Toward a Right to Exclude Migrants

Is there a moral right to exclude strangers? Many people around the world have an interest in crossing the borders into other countries. The reasons for their migration are various, but the most general motive is the hope to improve one's situation, whether economically, socially, or politically. The number of worldwide migrants is already high and likely to rise in the years and decades to come; at any rate, their existence forces potential host countries to make a decision on whether or not to exclude them. Legally speaking, this decision is at their discretion to make as sovereign states. But there is also a moral issue. For immigration concerns human interests, hopes, and needs, and it potentially touches on fundamental human liberties. So whatever the legal norms and conventions may be, they should be informed by that which is specifically moral about immigration.

In this chapter, central aspects of the morality of immigration shall be discussed.[1] After rejecting the view that there is a (human) right to global free movement, I will make a case for the contrary view that the exclusion of migrants is in principle justified. As John Rawls has it, "justification is a matter of the mutual support of many considerations, of everything fitting together into one coherent view."[2] It is in this vein that I will in this chapter bring together different argumentative strands that together converge toward the view that states have a *prima facie* right to exclude migrants. This right will be presented as a self-limited one in the sense that there may be conflicting considerations. The applicability of the right to exclude would appear to be compromised (that is, the legitimacy of *exerting* that right could be compromised) if in

---

[1] Note, however, that this chapter does not aim at providing a complete account of the current debates in the ethics of immigration. For comprehensive accounts of the ethics of immigration, see e.g. Carens (2013), Wellman and Cole (2011), and Miller (2016).
[2] Rawls (1971, p. 21).

a *particular* case or context the exclusion of (particular) people turned out to be altogether inconsistent with the assumption of those excluded people's equal basic rights.[3] Note that my case for a right to exclude has two interrelated strands. While the elaborations on the relevance of culture, common identities, and social trust (in 3.2) will culminate in my case for a *right to protect* (3.3), my support for Ryan Pevnick's *associative ownership view* (in sub-chapter 3.4) and my ongoing emphasis on the importance of self-determination will substantiate my claim that there is a *right to control* immigration.[4] Each strand will be integrated systematically into the previously established view of states as dividing moral labour (MDL). Together, these two strands will allow me to solidify the view that a *right to exclude* is, as of today, indispensable for preserving the sphere of freedom and security that states are to provide for. However, as already suggested, I will point out at the end of this chapter that there can be important qualifications to this right to exclude. There are circumstances under which the exertion of the right to exclude will be very dubious indeed.

3.1. Rejecting the Cosmopolitan Case for Global Free Movement

Is there in principle a right to global free movement? Defenders of that view conceive of open borders as the normative standard,[5] and they hold that even under most real-world constellations it would be wrong to hinder a migrant from immigrating into another country. In this first sub-chapter, I will challenge this case for open borders. For that purpose, I will exemplify and discuss the open borders view by drawing particularly on the position of Joseph Carens, who is presumably the most widely cited proponent of global free movement. Needless to say, authors other than Carens (who oppose or share his views) will also be included in that discussion so that after all I will

---

[3] I owe these two qualifications to Pevnick (2007).
[4] For a helpful discussion of the distinction between arguments for a right to protect and arguments for a right to control and for an assessment of how these two types of arguing for exclusion are used in current debates on the morality of immigration, see Cassee (2016, pp. 31-37). Note that the right to protect is in principle contingent on the existence of something that requires protection and that is worthy of being protected. The right to control, on the other hand, is not contingent in this sense but is an essential part of what it means to be self-determined.
[5] Michael Dummett contends that any deviation from that standard of open borders stands in need of justification. In his words: "The onus of proof always lies with a claim to a right to exclude would-be immigrants" (Dummett, 2001, p. 57).

come to discuss a whole range of distinct arguments and argumentative strategies in favour of open borders. In so far as the here discussed views and rationales are exemplary of the wider group of authors in the literature who argue in favour of (more) open borders, my critique of these discussed views and rationales will, at least to some extent, apply to that broader group of authors as well.

3.1.1. Why Global Inequality Does Not Justify Free Movement

A first argument that Carens presents in favour of open borders draws on liberalism's central tenet of equality, and more precisely on the equality of opportunities. As Carens observes, there is a large wealth gap between today's countries and therewith vast inequalities concerning the opportunities that people find in their respective countries. A person who happens to be born in a poor country will have access to a much smaller range of opportunities for leading a good life than another person who happens to be born in a more prosperous country. Famously, Carens compares this circumstance with the institution of feudalism.

> Since the range of opportunities varies so greatly among states, this means that in our world, as in feudalism, the social circumstances of one's birth largely determine one's opportunities. It also means that restrictions on freedom of movement are an essential element in maintaining this arrangement, that is, in limiting the opportunities of people with talents and motivations but the wrong social circumstances of birth.[6]

Departing from this diagnosis, Carens conceives of global free movement as an important moral goal because it would, in his view, decrease the degree of global inequality that he thinks is brought about by the way the world is currently organized. As I will contend now, this line of argument must be rejected for several reasons. To begin with, the analogy between feudalism and a world of states separated by borders is flawed. The problem with this analogy is that other than in the case of feudalism, which used to be disadvantageous to almost anyone living under the feudal regime except for a small ruling class, today's world's border regime is not that unambiguously detrimental to the vast majority. At least in functioning states—which obviously need not be a minority of the total of states in the world—*most* people will have a legitimate interest in borders. Borders will, for one thing, allow citizens to *control* foreign people's access to certain goods which they (more that others) brought about through their cooperative efforts

---

[6] Carens (2013, p. 228).

so that they not only hold an ownership claim over those goods but also appear to have a *prima facie* right to determine how and by whom such goods are used. For another thing, borders will allow them to *protect* certain goods and conditions on which their cooperation depends.

Even those people in developing countries who as of today have relatively limited opportunities would — given improved economic conditions in their own countries — endorse borders as well, on the same grounds that (most) citizens of other already successful states endorse borders. And even before economic conditions improve, the people living in, say, the Philippines or Morocco, i.e. people who by Carens' worldview figure as victims of an alleged global feudalism, will presumably have an interest at least in their *own* countries having borders, in controlling access to the goods they created and in protecting certain (cultural) conditions that they value, poor though they may be. That rights require community, and that communities require borders — as I will argue more systematically in following parts — is true for all countries, be they presently successful or not. As regards the case of feudalism, this is not true at all. People will reject feudal arrangements at all times and places. Carens analogy thusly fails.

There is another reason why his equality-based argument for open borders fails to be persuasive. To begin with, one may readily concede that the current set-up of the world hinders a large group of people from grasping all the opportunities available, and there is therefore of course a sense in which people are left with dramatically unequal opportunities to achieve a certain degree of material success in life. But the problem I am alluding to here lies in the way in which Carens links up the empirical observations on globally unequal opportunities on the one hand with normative demands concerning the existence of (closed) borders on the other hand. On the face of it, his reasoning appears to be straightforward: he observes that one group of countries provides its citizens with more attractive a range of opportunities than the other group, which he judges to be an unfairness, and he then departs from this assessment to contend that people from the worse-off countries should be allowed to move freely to the better-off countries so as to reduce the presumed unfairness. But this reasoning is problematic. A first reason why this is so is that it neglects the possibility that the broader range of opportunities available in a given country A is — among other factors — the result *precisely* of its ability to exclude others from its territory.

Take the hypothetical example of Japan. If Japan's present day prosperity was made possible or facilitated (among other factors) by its right to exclude and thusly to protect itself economically and culturally, then it would present a doubtful argumentative move to call for open borders in order to have other (non-Japanese) people share in this prosperity. To the extent that the broader range of opportunities (or prosperity) would not exist or would not have come into being (in any form comparable to the current presumed prosperity) without Japan's right in the past to fence of the intrusion of foreign people and influences, and in so far as the exclusion of foreign people is not *intrinsically* wrong, which at least the present argument neither presumes nor aims to establish, it would be short-sighted if not absurd to call for a measure (namely open borders) which denies Japanese people that authority (namely to exclude). Together with other factors, it is the authority to exclude that helped bring into existence the now envied thing (i.e. Japanese prosperity) in the first place.[7]

Japanese people will find the reasoning of authors like Carens disconcerting. When they have to defend themselves against such cosmopolitans who on moral grounds deny the Japanese right to decide on who can become a Japanese citizen, the Japanese could further point to the generally beneficial effects of a system of (moral) division of labour in which self-responsible states strive for sustainable levels of prosperity. They could readily concede that the current way of organizing this system of states is marked by blatant injustices. And they might concede that the deeper moral problem with existing inequalities is that they tend to give rise to political inequalities as well, which in turn allow the richer and more powerful states in that system to exploit and oppress the weaker ones.[8] But rather than giving in to Carens' call for global free movement, countries asked to waive their right to exclude could point to the existence of alternative and indeed more sustainable (non-immigration related) measures that would likely ease the current discrepancy between the ranges of opportunities available to people born in different countries.[9] One could push for a restructuring of

---

[7] For support of the modest thesis that a country's prosperity depends (among other things) on the effectiveness of its institutions and that the immigration of people from countries with dysfunctional social models can have a negative effect on that effectiveness, see Collier (2013, pp. 28–37).

[8] See Beitz (1979, pp. 146–50) and Anderson (1999, p. 314).

[9] That large-scale immigration may, at least in an imperfect world, not be the most sustainable (that is, stability-enhancing) measure for a (better-off) country to combat global inequality is underpinned by Fareed Zakaria, who by drawing on a comparison between Japan and Western countries points to the relationship between immigration and populism's rise: "In 2015, there were around 250 million international migrants and 65 million forcibly displaced people worldwide. Europe has received the largest share, 76 million immigrants, and it is the continent with the greatest anxiety. That anxiety is proving a better guide

international trade, or one could work toward an increase in the degree of global redistribution of wealth. Whatever duties rich countries may have toward globally disadvantaged countries (under distributive justice), it is not clear why such duties should be cashed out in terms of open borders rather than in any other form.[10] In light of alternative institutional measures and policies for reaching a more just distribution of global wealth, the focus on calling for open borders is hasty, one-sided, and risky. As moral psychologist Joshua Greene reminds us, the maintenance of successful cooperation remains "our greatest challenge".[11] To the extent that human brains are still "wired for tribalism" and in so far as humans still "intuitively divide the world into Us and Them, and favor Us over Them",[12] it appears that a world of open borders for the sake of reducing inequality will only make this great challenge of maintaining cooperation more difficult. This suggests that Carens' call for open borders fails to be persuasive even in a world that *is* characterized by a shameful degree of inequality across countries.

Beside the mentioned hazards on the side of the host society, the possible dangers for the societies of source countries should not be neglected. As David Miller remarks with regard to the already mentioned risk of "brain drain":

> Indeed, a policy of open migration may make such people worse off still, if it allows doctors, engineers, and other professionals to move from economically undeveloped to economically developed societies in search of higher incomes, thereby depriving their countries of origin of vital skills. Equalizing opportunity for the few may diminish opportunities for the many.[13]

In light of the risks that an open borders policy means for source countries and destination countries alike and in light of the fact that open borders are only *one* possible means of easing the current degree of global inequality, to open the borders appears to be an *unnecessarily*

---

to voters' choices than issues such as inequality or slow growth. As a counter-example, consider Japan. The country has had 25 years of sluggish growth and is aging even faster than others, but it doesn't have many immigrants—and in part as a result, it has not caught the populist fever" (Zakaria, 2016, p. 14).

[10] See Wellman and Cole (2011, pp. 65-66).
[11] Greene (2013, p. 59).
[12] *Ibid.*, p. 54.
[13] Miller (2005, pp. 198-99); similarly quoted in Wellman and Cole (2011, p. 68). As Miller further points out, what makes "brain drain" even more problematic is that the poorest members of source countries will often lack the means to migrate to wealthier countries even in the event of open borders. It appears to be them who suffer most from the effects of "brain drain" (Miller, 2005, pp. 198-99).

risky measure. This assessment leaves untouched the notion that richer states have a strong moral responsibility to reduce the current inacceptable level of global inequality.

3.1.2. Why Global Free Movement Is Not a Human Right

The call for open borders as discussed in the previous section was contingent on the reality of an unequal distribution of opportunities across countries. After that line of argument was rejected, it now bears looking at a more *principled* way of arguing for global free movement. Couldn't one argue that the freedom to move across borders and to choose a place of living wherever on Earth a person wants to amounts to an essential human interest so that one may declare this interest the substance of a *human right*? Joseph Carens makes this claim, and the aim of this section is to examine and eventually to reject that claim and the central argument he adduces to support it.

The argument I will focus on in this section is referred to by Carens as the "Cantilever Argument". According to Carens' characterization, the cantilever strategy means that one makes "a normative argument in favor of recognizing a new human right" by showing that this new human right "is closely analogous to something that we already recognize as a human right".[14] Now, in order to argue for a right to global free movement, he makes use of this cantilever strategy in the following way: he first assumes it to be a human right for people to move freely *within their own country*, and then he contends that—as a "logical extension"[15] of this right to internal free movement—there must also be a human right to move freely across international borders:[16]

> Every reason why one might want to move within a state may also be a reason for moving between states. One might want a job; one might fall in love with someone from another country; one might belong to a religion that has few adherents in one's native state and many in another; one might wish to pursue cultural opportunities that are only available in another land.[17]

Based on this series of assumptions, Carens concludes that "[t]he radical disjuncture that treats freedom of movement within the state as a human right while granting states discretionary control over freedom of movement across state borders makes no moral sense."[18]

---

[14] Carens (2013, p. 238).
[15] *Ibid.*
[16] A similar line of reasoning is found in Oberman (2016).
[17] Carens (2013, p. 239).
[18] *Ibid.* Indeed, if one did not give much weight to the consideration that the members of a particular room of justice have a *special right* to the goods they

In the following, this reasoning of Carens will be criticized as misguided. I will make two interrelated objections. In a first step (I), I will argue that neither internal nor international free movement truly qualifies as a human right. So rather than criticizing only the way in which he tries to expand the scope of the right to free movement from the national to the international level, I will go a decisive step further and already call into doubt (somewhat provocatively) that *internal* free movement qualifies as a *human right* (at least I will question that it qualifies as a human right in the strict sense, which as we will see does not preclude that people will be standardly entitled to internal free movement). In the second step (II) of my critique of Carens I will be concerned with his contention that every "reason why one might want to move within a state may also be a reason for moving between states".[19] Here I will object that there clearly *are* fundamental (non-human rights-related) reasons that speak in favour of internal free movement but which simply do not apply to global free movement. This will finally make plausible what Carens is unwilling to accept: namely the existence of a "radical disjuncture" between (free) internal movement and (restricted) international movement.

I. Getting Human Rights Straight

Concerning my first line of argument against Carens' cantilever argument, it is worth reflecting on the nature of basic human rights. As the feature of central concern here, recall that human rights designate a *moral minimum*. In Gewirth's terms, they designate a minimum in so far

---

created through their cooperative efforts, then there would be little that stands in the way of declaring these goods the substances of human rights and then of extending the right to immigrate to all people in the world because it would be only by immigrating (i.e. by being granted a right to immigrate) that all those people could access those things. Andreas Cassee does not stop here though. In his view, the exertion of the right to global free movement triggers "a whole series of further demands" (Cassee, 2016, p. 214, my translation). Those who decide to exercise their right to move to any other country would there have, among other things, the same rights to political participation and to social welfare as all other citizens of that country. As Ott criticizes, the final demand of many proponents of open borders is then not only open borders, but open borders in combination with equal rights to social welfare and other valuable goods for all people in the world (Ott, 2016, p. 74). For Carens, this appears to be justified to avoid an unjustified defense of privilege (Carens, 1988, p. 217). To the critical reader interested in practical moral guidance in the realm of immigration policy, such calls for free global movement and social welfare will rather seem like an absurdity.

[19] Carens (2013, p. 239).

as their contents (or their substances) are the necessary goods for free and purposeful action and for leading a minimally decent life.[20] Or in Shue's terms, they designate a minimum in the sense that "any attempt to enjoy any other right by sacrificing the basic right would be quite literally self-defeating, cutting the ground from beneath oneself."[21] The decisive point to be made here and that I will presently use in order to contradict Carens' notion that free movement should be seen as the object of a human right is simply this: not to have (socially safeguarded) access to the substances of a basic right—i.e. not to enjoy those substances as rights—leaves the individual in a most vulnerable position. More still: in those moments when she effectively lacks those substances, she will be left in a most *degrading* situation, for she finds herself deprived of the very ability of free and purposeful action. She then lacks the means for leading her life in a minimally decent way.

This characterization makes available to us a way of testing whether something qualifies as the object of a human right. The test consists in asking whether the deprivation of the substance in question will leave the bearer of the putative human right in a degrading situation, i.e. one in which his ability to lead a minimally decent life will be seriously compromised. Now, regarding Carens' normative assessment that global free movement should be considered a human right, it would be bizarre to assert that a person is indeed degraded and deprived of genuine freedom of action when she is denied the right to move freely across the globe. David Miller shares this scepticism when he calls into doubt the portrayal of *global* free movement as an *indispensable* requirement for genuine freedom of action. What Miller argues instead is that only the freedom to move within a *sufficiently large* territory would qualify as a human right:[22] it is certainly the substance of a human right that one not be hindered from moving over "a fairly wide area"[23] in order to find a job, to meet a partner, etc. But to declare *global* free movement the object of a human right clearly misses the very point of what the normative concept of human rights is meant to bring home: namely that to have certain basic goods protected against standard threats is absolutely essential for *leading* a minimally decent life. In a

---

[20] As Gewirth asks trenchantly in this respect: "If a rational agent is to claim any rights at all, could anything be a more urgent object of his claim than the necessary conditions of his engaging both in action in general and in successful action?" (Gewirth, 1978, p. 63).

[21] Shue (1996, p. 19).

[22] See Miller (2005, pp. 194-96). For this line of sufficiency-based objection to Carens see also Wellman (2016, pp. 87-90), or Pevnick (2011, pp. 84-86).

[23] Miller (2005, p. 195).

nutshell then, if something is not necessary for fulfilling one's basic needs, it cannot be the substance of a basic right, and global free movement is no such thing. It may be considered good and freedom-enhancing for some people to be allowed to move freely across international borders, but as Jack Donnelly has it: "Most good things simply are not the object of human rights."[24]

Indeed, by this understanding, even *internal* movement is not a human right. The *movement as such* is not basic in the sense that it is necessary for leading a minimally decent life. From the standpoint of basic rights, the freedom to move all across Germany, Argentina, or Russia is just as dispensable as the freedom to move all across the globe. As Ryan Pevnick reminds us, a decent life would not be possible when permanently confined to a single room, and there are presumably few who "would deny that restriction of this type amounts to a rights violation".[25] But there is no reason for assuming that by the above account of what characterizes human rights, the freedom of internal movement is indisputably the content of such human rights. After all, the fact that free internal movement is codified as a human right in international legal documents does not compromise the assessment that from a purely morally-normative standpoint it makes nonsense of the idea of basic rights to contend that a Canadian denied the freedom to move from Vancouver to Québec will find it difficult to enjoy a minimally decent life. (I hasten to add though that I will presently argue for another reason why every citizen will invariably enjoy—as a right—the freedom to move freely on the territory of his or her own state and, to be sure, this right is *derivative of* a genuine human right.)

As a more fundamental critique of Carens' cantilever argument, I will now argue that the reason why citizens indeed have a right to move freely within their own states is not that such a freedom is important or that citizens have an especially deep interest in it. Once it is established that the basis of the notion that citizens are entitled to free internal movement is not their *interest* in such movement, Carens' argument that non-citizens must equally have a right to move freely within that (and any other) country because they equally have a strong interest in doing so will lose its normative thrust. Once the focus shifts away from interests, it will turn out that the reasons why citizens have a right to free movement within their country simply do not apply to non-citizens.

---

[24] Donnelly (2013, p. 17).
[25] Pevnick (2011, p. 84).

II. The Disjuncture between Internal and International Movement

As Carens sees it, the same reasons that speak in favour of a right to free internal movement also speak in favour of a right to international free movement, "because the reasons why people *want* [my emphasis, J.K.] to move from one place to another will apply in both cases".[26] The logic behind this reasoning is questionable. As a general observation that calls into doubt this reasoning of Carens, note that while you can want all kinds of things, what you want is often not the decisive thing. The decisive thing is what you have a *right* to. And what you *want* will often have no effect on what you have a *right* to. You can, for some reasons, want one thing very much and you get it because you have a right to it, and then there may be another relevantly similar thing that you also want very much for the same reasons but which you do not get, because you have no right to it. For example, it may be so that for several reasons you want a material minimum of social protection in and by your own state and that you indeed have a right to that minimum, but even though you may have the same (good) reasons for wanting such social protection in another country, you have no right to social protection in that country. The mere fact that you want it has—in this case—no implication for the question of whether or not you have a right to it. Put somewhat pointedly, Carens' logic appears to be that if only some people want to move internationally just as much as they want to move internally they must have the right to do the former in just the same way that they have a right to do the latter. And it is this kind of reasoning that leads him to drawing the wrong conclusion in the case of whether or not there is a right to free movement across international borders in just the same way (or for the same reasons) that there is a right to internal free movement. In fact, there is a disjuncture between *wanting* something and being in a position to claiming that wanted thing as a *right*. It is this disjuncture that explains and justifies the "radical disjuncture" between a (presumed) right to internal movement and the inexistence of such a right at the global level. After all then, that (some) people want or even need to be allowed to move into a foreign country does not confer on them a right to do so. But this verdict leaves unanswered the decisive remaining question. What is it then, if not volition or need, that confers on some (i.e. the citizens) but not on others (i.e. all non-citizens) the right to enter and move freely within a given country?

Arguably, and as I will contend more systematically in the following sub-chapters in which I develop a positive case for a right to

---

[26] Carens (2013, p. 239).

exclude, the citizens of a state—who in their entirety make up for and *are* the state—*own* the state and its institutions in the simple sense that they rather than others contributed to and created those institutions through their cooperative efforts. How could it be justified to a citizen that he standardly may not use the state's highways—which he helped build through his tax contributions—in order to travel or move to the other side of the country? The only way of arguing that there could be such a general justification would begin by assuming a most doubtful dichotomy between *citizens* and *state*, and based on this dichotomy one could then contend that the state has a right to standardly deny the citizens this right to move freely within the state's borders. But in so far as the citizens of a democratic state in an important sense constitute and *are* the state, it is far from clear why they should accept that their own state denies them this freedom. In democratic states there can be no agency that standardly denies citizens to move freely across what is theirs. The citizens of a democratic state naturally have the right to internal free movement then simply in so far as no one can rightfully deny them the freedom to access the goods they can call their own. Non-citizens, on the contrary, cannot point to this ownership structure. They cannot claim to have contributed through their taxes to the infrastructure (roads, hospitals, schools, etc.) of the country in question. In their case, there *does exist* an authority that can deny them to move freely in that country: namely the democratically constituted group of citizens who contributed to the existent (institutional) infrastructure and who therefore have a (*prima facie*) right to control non-contributors' access to that infrastructure. The very least that these ownership-based considerations suggest is that there is some relevant difference between citizens and non-citizens when it comes to the question of why one but not the other group should be entitled to internal free movement. Carens may be right that the same reasons why citizens *want* to move within their country will equally apply to non-citizens. But Carens misses the mark here: non-citizens are simply not equally *entitled* to such free internal movement.

In fact, the only (unlikely) reason why free internal movement could be standardly denied even to citizens appears to be that they *themselves* decide—democratically—that they not be free to move on the territory to which they can claim ownership.[27] While this democratic self-

---

[27] I hasten to add that even if they were to decide this, this democratic decision would be limited by a human rights concern as already raised before: namely that some degree of freedom of movement is a basic right that must be secured in order to make sure that people are minimally free to lead their lives in the dispositional sense.

restraint may be considered a rather unlikely scenario, the focus on the relevance of the citizens' *decision* leads to a related argument in favour of the view that citizens have a right to move freely within their own country: as a democratically constituted people, they have the right to decide freely on how they organize their communal life and how they regulate access to their own assets like their territory, and if they *decide* that they be allowed to move freely on that territory, then they have this *right* simply in so far as they decided to have it. Citizens clearly have a *human right* to participate democratically in decisions that regulate the way they live together in their political community, and to the extent that they likely will not use that human right so as to restrict themselves in moving freely on their political community's territory, it may be said that citizens' entitlement to free internal movement is derived from their human right to self-determination over how internal matters are regulated. Their likely decision not to restrict internal movement will maximize their own freedom, and to the extent that this enhanced freedom is a core reason why citizens want and need their state in the first place, it would be an absurdity to hold that the state has a right (i.e. that the citizens will decide) to deny the freedom to move internally. By the here proposed rationale in favour of free internal movement, there remains indeed no reason to assume that non-citizens *equally* have a right to internal free movement.[28]

---

[28] After all, *even if* one accepted the view that there is a human right of non-citizens to immigrate into a given country, there would — presumably — still remain situations in which that country would be well-advised and morally right not to actually allow certain non-citizens to immigrate. As Henry Shue recognizes, "sometimes, rights must be violated — or, at the very best, violations of rights must be overlooked" (Shue, 1983, p. 10). If in particular (extreme) situations it were *counterproductive* or even self-defeating to grant people such a (presumed) human right to immigrate, for example when granting that human right would risk the stability and effectiveness of the institutions necessary for the protection of human rights (of citizens) in the first place, then the exclusion of migrants could well be justified. The infringement of their presumed human right would be proportional and justified because and in so far as the contribution of this infringement to the aim of maintaining public order (and thus of retaining the integrity of the citizens' "room of justice") in which human rights are secured would appear to be "greater than the harm done by the violation" (Shue, 1983, p. 11). In fairness, it must be said that Carens acknowledges the possibility that global free movement would sometimes have to be constrained due to competing considerations, for example due to a concern with public order (see e.g. Carens, 2013, pp. 276–78). But he fails to acknowledge just *how much* such competing concerns could in actual fact limit the applicability of the right to free international movement as he postulates it. Presumably, it is only by couching the concern with public order as a risk-ethical one that one can

3.1.3. Open Borders and the Disregard for the Value of Self-Determination

Having already provided over the course of this sub-chapter a series of moral arguments against global free movement, I will now reject this call for open borders by pointing out how a right to global free movement would severely limit the possibility for political communities to be meaningfully self-determined.

My following argument will repeatedly draw on the example of Iceland. This country—being rather small both in terms of its territory and in terms of its population size of only around 300,000—presumably needs only a *relatively* low number of immigrants until first societal, environmental, or political changes can be perceived. This makes it a suitable test case for asking the question of how deeply the absence of a right to control immigration (with some discretion) is in tension not only with our conceptions of collective self-determination but, by extension, also with individual freedom. After all, the following reflections on Iceland aim at stimulating the reader's imaginative power: how would the right to global free movement bear on Icelanders' ability to pursue their current conceptions of the good life?

Let me begin with a few seemingly random notes on Icelandic society. Among OECD countries, Iceland ranks at the top in several areas such as environmental quality, safety (i.e. a remarkably low murder rate), and job security. It has the best employment rate of all OECD countries, and there is no other OECD country in which—in terms of income—social inequality is as low as in Iceland.[29] Moreover, there is "a strong sense of community and high levels of civic participation".[30] As such findings indicate, Icelanders can pride themselves with several morally valuable goods, most notably their achievements in terms of equality, security, and individual freedom. Importantly, however, one should not assume such achievements to be natural facts of life. They are the result of a concrete conception of how to organize communal life that prevails among (current) members of a concrete polis. In Paul Collier's technical words, such achievements are expressions of a particular "social model" whose maintenance is facilitated by the prevalence of certain norms and institutions within Icelandic society.[31] As a concrete polis on a defined territory, Icelanders had the possibility to foster and realize their conception of justice

---

fully grasp just how much such competing considerations could compromise the applicability of his open-borders argument. But Carens assumes no such risk-ethical perspective.

[29] See OECD (2015) and OECD (2013).
[30] OECD (2015).
[31] See Collier (2013, pp. 33–37).

within a room of justice.[32] Iceland is their room of justice. But could it be that without walls?

Any social model can come under pressure through certain societal developments and, as my earlier empirical account of the effects of immigration suggests, one possible source of such developments can be certain forms immigration. This is particularly likely when such immigration occurs in response to a radical open borders policy. As already argued at earlier stages, Carens does not take the danger of such a policy seriously though. He does not take seriously how exactly an open doors policy could bear on concrete exemplary countries like Iceland. When immigration occurs, it is real people that come – people with their own individual histories and mindsets. While this rather obvious observation is far from trivial, it is Carens' who trivializes the potentially negative impacts of immigration when he notes that the immigrants who make it to another country are generally "ordinary, peaceful people, seeking only the opportunity to build decent, secure lives for themselves and their family".[33] Konrad Ott rejects this characterization of Carens as overly idyllic and as blind to the general truth that many migrants have been subject to radically different local processes of socialization. If cultural-philosophical discourses on the otherness of people from other cultures are taken only minimally seriously, then, according to Ott, one cannot but assume that the people who come will bring with them their convictions and religious beliefs, their codes of clan and understanding of family honour, their desires and traumata, and their loyalties and feuds.[34] As Collier notes in a similar vein, "[m]igrants bring not only the human capital generated in their own societies; they also bring the moral codes of their own societies."[35]

This could mean that in particular cases immigration might turn out to be less enriching for society than generally asserted; it risks putting a serious strain on the institutions and achievements of exemplary countries like Iceland. There can be no doubt that social, political, and economic interaction is structured by such informal constraints as customs, traditions, taboos, and codes of conduct more generally,[36] and if these elements change through certain forms of immigration, then

---

[32] See Sitter-Liver (2003, p. 40).
[33] Carens (2013, p. 225).
[34] See Ott (2016a, p. 74) and Collier (2013, p. 32).
[35] Collier (2013, p. 67).
[36] See North (1991) and his broad conceptualization of institutions as both *informal* constraints such as customs, codes of conduct, etc. and *formal* constraints such as laws, constitutions, or organizations.

this cannot but have an effect on forms or even the functionalities of societal interactions. When the migrants who join the current citizens bring with them archaic social models, then the current citizens appear to have a good reason to be worried about such forms of immigration, and it is only commonsensical to assume that the import of such archaic social models will lead to inner-societal clashes of one sort or another.

But such problems will remain hidden to those moral theorists who, in their philosophical reflections on immigration, prefer to remain at some distance from the ground at which such tensions occurs. Maybe unintentionally, Andreas Cassee admits this distance when he notes in the preface to his book on immigration that most of his views had to undergo a plausibility test on the balcony of a colleague in Zurich.[37] This reference to his reasoning on the balcony is telling. It attunes his reader to a normative perspective on immigration policy that may well be described as "Oxford armchair style".[38] It foreshadows an aloofness that allows him to brush aside the real problems and worries associated with immigration. The potential erosion of shared (national) identities through immigration plays a negligible role in his reasoning. He simply calls into doubt that such identities exist, let alone that they are relevantly of significance for citizens.[39] Most such citizens on the ground would of course deny that. After all, it appears that from the comfortable distance he enjoys on the balcony, both existing citizens and migrants easily pass as abstract entities whose identities, attachments, and values are indiscernible. This abstraction allows him to develop his case for a world with open borders. Clearly, however, the story this distance allows him to tell will cause unease for most people on the ground.

According to a more realistic story migrants are often "escaping from countries with dysfunctional social models".[40] And in so far as they bring with them their social models, it is hard to see how countries that have no (discretionary) right to control immigration should guarantee the prevalence of their own social model. This limited control over the prevalent social model implies a lack of control over the prevailing conception of social justice. And in so far as the core of what it means for a political community to be self-determined is that — inside a territorially defined room of justice — it can work toward the realization of a certain conception of justice, then once this association of people no longer has far-reaching and effective control over the con-

---

[37] Cassee (2016, p. 10).
[38] MacIntyre (1981, p. v).
[39] Cassee (2016, pp. 152–56).
[40] Collier (2013, p. 34).

tents of this conception of justice the whole idea of what it means to be self-determined is undermined. For if the underlying conception of justice is beyond control, then the realization of certain rights and norms that flow from the currently prevailing conception of justice is also beyond control, and the purpose of the room of justice is undermined. If the Icelandic state and its people have no far-reaching control over *who* they are and *how many* they are on their island, it is not clear in what sense they really control the maintenance and future course of their room of justice.

Without control over its own membership (i.e. the self), a political group cannot be called *self-determined* for almost tautological reasons, because in order to be self-determined it must be able to *determine the self*.[41] The *room* of justice would end up as *unbounded* in that case, and this would amount to a contradiction in terms. More to the point, the very idea of referring to a political community as a *room* of justice is that, by its very nature, a room makes it possible to capture and shield certain things (like a conception of justice and the rights it allows for) from outside influences. When cosmopolitans call for (fairly) open borders, they appear to declare such rooms unnecessary. It seems there are two possible explanations why one could consider such rooms unnecessary: either (1) because one regards the current way of organizing society within and via that room as not sufficiently worthy of being shielded, or because (2) one does not regard outside influences as a threat to that way of organizing society. As regards cosmopolitan reasoning on immigration, it is presumably mainly (2) that applies. My elaborations in this section suggest that the proponents of open borders had better re-examine this assessment (2). So far, they fail to grasp fully the risks that free immigration means to successful social models like the Icelandic one.

3.2. The Importance of Community and the Need for Exclusion

Community is important, as a primary value for most people, as a key determinant of happiness, and as a source of material benefits.[42]

---

[41] In a similar vein, Wellman observes that "an essential part of group self-determination is exercising control over what the 'self' is" (Wellman, 2013, p. 41). For him, "it seems clear that one cannot limit freedom of association without restricting self-determination", and if self-determination is considered morally valuable, it appears that "we should begin with a presumption in favor of freedom of association" (*ibid.*, p. 34).

[42] Collier (2013, p. 234).

That communities are important is a verdict few moral philosophers would deny. It is *communitarian* theorists, however, that give extra normative weight to that verdict. Before I unfold in the course of this sub-chapter my own case for the importance of community and its implications for normative reasoning on immigration, I will shortly characterize that communitarian appraisal of community, and I will criticize how it results in unacceptable views on immigration policy. This rejection of the communitarian take on immigration is necessary not only because communitarian theorists often take centre stage in discussions in the ethics on immigration. Just as importantly, it will also serve as an important touchstone for the discussions in subsequent sections on the *proper* role of culture and the role of identity and social trust. After all, what this sub-chapter aims to establish is the view that community is something that is worthy of protection. If this modest view can be established, the way for arguing in favour of a *right to protect* in the next sub-chapter will be paved.

3.2.1. The Communitarian Overemphasis on Community and the Problem with Exclusive Identities

Communitarian theory emphasizes that community describes not only what citizens *have*, but also what they *are*, and it is argued that a member's commitments to other members forms the basic element of personality.[43] Community figures in such accounts as more than just the sum of its parts. It is rather a moral value in its own right. By that view, one can have special duties to co-nationals for no other reason than their being co-nationals.

Regarding the implications of this assessment of community for the issue of immigration policy, it bears taking a look at the reasoning of Michael Walzer, a communitarian theorist who is widely quoted in current debates on the morality of immigration policy. As Walzer contends, communities will by their very nature (cohesion) be repugnant to open borders. Drawing on his analogy between neighbourhoods and borderless states, he argues that "if states ever become large neighborhoods, it is likely that neighborhoods will become little states."[44] Or, as

---

[43] Sandel (1992, pp. 20–24) and Miller (2007, p. 238).
[44] Walzer (1983, p. 39). Walzer invites the reader to imagine a world in which political communities are like neighbourhoods, i.e. a world in which countries have no authority to control their membership. He argues (by drawing centrally on passages of Henry Sidgwick) that if the labour market were the decisive force in determining where people live, the result would be associations of rather casual and fluctuating character. No patriotic sentiments would evolve in such associations, and with individuals constantly on the

he famously puts it in a similar vein: "To tear down the walls of the state is not, as Sidgwick worriedly suggested, to create a world without walls, but rather to create a thousand petty fortresses."[45]

Note that by the communitarian view, co-nationals share a common identity that draws on such features as a common history and culture, a common language, forms of life, and, on some accounts, also a common ethnic background. But being built on such features, it will remain an inherently *exclusive identity*.[46] It relies on features that non-citizens cannot (easily) adopt. This recognition finally allows me to turn to the doubtful take of communitarians on immigration policy.

In order to preserve what Walzer calls "communities of character", people who do not share a certain national (exclusive) identity can be excluded simply on the basis that they do not share it. Insiders could argue that this identity would come at risk through the (numerous) immigration of such people who do not portray the relevant features of that exclusive identity. (And indeed, as already indicated, they will hardly ever be able to portray such features.) In this sense, by communitarian reasoning, it would generally be argued that political communities have a *discretionary* right to exclude potential immigrants. All they need to rely on for that claim is the alleged importance of preserving the distinctive *character* of the community. Without this right to control membership, Walzer fears, no "communities of character" would be possible. There would be no shared understanding of a common good and there hence would be no special commitments among members.[47]

To allow a state to exclude people based on such an *exclusive* understanding of culture provides that state with a rationale that it could make use of in just too arbitrary a way. What makes a common identity especially *exclusive* is when it is anchored in such features as race or accent-free command of a language. Where this understanding of identity prevails, the doors for arbitrary misuse stand wide open and some people will never get even the chance to immigrate. Moreover, if a particular group were denied membership on the grounds of some

---

move there would be no cohesion among members of that association. People would end up as strangers rather than as neighbours to one another. Without the possibility of maintaining a certain degree of cultural homogeneity among citizens it would in Walzer's view be difficult to promote a moral and intellectual culture. Moreover, the establishment of efficient political institutions would be impeded (*ibid.*, pp. 37–39).

[45] Walzer (1983, p. 38).
[46] See Kymlicka (2001, pp. 258–62).
[47] Walzer (1983, pp. 61–62).

feature they portray, then this would imply a striking discrimination of all those existing citizens who already portray that very feature, which in today's Western and pluralistic societies will regularly be the case. This discrimination risks being racist and is therefore unacceptable. The more the focus is laid on the particular *character* of a community at a given point in time, the more exclusionary the immigration policy will likely be. In fact, the deeper problem with the communitarian take on immigration policy is twofold: (1) that some individuals have by chance of birth no chance of becoming "true" members, and (2) that rejected individuals get no justification for their exclusion other than that existing members of the community judge them to be unfitting and that through their admission the "character" of the community would be at risk of being altered too much. This lack of justification is unacceptable.

To give exclusive identities such a central place in moral reasoning — and indeed to consider particular cultural contexts as the only meaningful starting point and endpoint of normative theorizing on immigration policy — brings along a more fundamental problem. As liberal philosopher Onora O'Neill convincingly suggests, "[a]n attempt to think of actual boundaries as horizons that are constitutive of identities and understandings can only disorient all discourse about global concerns."[48] These reservations of O'Neill amount to an important reason why one should be sceptical of the communitarian take on immigration. It tends to *overemphasize* the normative power of identity in a world where many other relevant considerations play a role. The goal of preserving the "character of a community", though not necessarily illegitimate in itself, is unduly taken as the primary and ultimately decisive consideration. "Those who see boundaries as the limits of justifiable reasoning will not take seriously — indeed may not be able to acknowledge — either the predicaments of those who are excluded or the alternatives for those who have been included."[49]

And yet there seems to be something in communitarian reasoning that meets with our very intuitions. Membership in a community matters to most of us. We do normally value community and culture and we find such values somehow deeply entrenched in our identities. I suggest that culture should play an important role in any account on immigration — even in approaches as the present one that is concerned with free and equal individuals, not with culturally bounded members of communities. But recall that it was already recognized in the second

---

[48] O'Neill (2000, p. 169).
[49] *Ibid.*

chapter that the freedom and well-being of individuals can only be realized through the mutuality and cooperation within communities. So what role do culture and identity play in allowing for that mutuality and cooperation within community? What is the proper place of culture and community within an account that takes the rights of all individuals equally seriously?

3.2.2. The Proper Role of Culture as a Context of Choice

> [I]t is the community that protects and fulfills these rights, especially for the persons who do not effectively have them, on the basis of the mutuality that characterizes human rights and the consequent solidarity of the society whose members are brought together by their recognition and fulfillment of common needs. Thus rights require community for their effectuation, and community requires rights as the basis of its justified operations and enactments.[50]

> The nation is primarily valuable not in and of itself, but rather because it provides the context within which we pursue the things which truly matter to us as individuals—our family, faith, vocation, pastimes, and projects. As Jonathan Glover puts it, a useful maxim for liberal nationalists is: "Always treat nations merely as means, and never as ends in themselves."[51]

What is the proper value that a rights-based approach should attribute to cultural membership? In his book *Liberalism, Community and Culture* (1989) Canadian philosopher Will Kymlicka offers an instructive account of the ways in which the categories of community and culture can and should enter liberal political philosophy. For the purposes of the present concern with immigration, his assessments will turn out to be very helpful—Kymlicka presents the key for understanding why culture and community, and the related concept of national identity, should play important roles alongside an unabated emphasis on individual rights and freedom. He offers the bridge between the intuitive appeal of communitarianism and the seemingly more individualistic rights-based approach applied here. The idea in this section is not to adopt Kymlicka's approach unaltered though, but rather to amend it in parts and apply it to the issue at stake here: immigration and the possibility of morally legitimate exclusion.

As such, culture does not have a systematic place within liberalism.[52] Its moral ontology recognizes only individuals.[53] So in order to

---

[50] Gewirth (1996, p. 97).
[51] Kymlicka (2001, p. 260).
[52] I will assume here that the rights-based approach I provided above is a variety of liberalism, quite simply in so far as it shares with liberalism its emphasis on

introduce culture in a meaningful way into liberalism—without fundamentally challenging the focus on individuals—it must be shown that, for the individual, membership in a given culture is an important good; that the individual not only belongs to the community but also that a particular community belongs to the individual in the sense of a mutual dependency. Kymlicka's central claim is that cultural membership is *essential* for securing the *individual* interest in leading a life that she judges to be good. If this can be argued for plausibly, there will be established a meaningful way of talking about and valuing culture, community, and identity from the standpoint of a rights-based ethic.

According to Kymlicka, a central claim of liberalism's political morality concerning the individual goes as follows:

> Our essential interest is in leading a good life, in having those things that a good life contains. That may be a pretty banal claim. But it has important consequences. For leading a good life is different from leading the life we currently believe to be good—that is, we recognize that we may be mistaken about the worth or value of what we are currently doing.[54]

It is for this reason that it is so important for the individual to be free. We should be free so that we can choose freely between the options available to us. We want to choose freely in accordance with our current conception of the good, which in turn we are free to revise or reject.[55] The crucial point concerning the present question of culture's role is that even where we decide freely on how to lead our lives, "the range of options is determined by our cultural heritage".[56] If we want to live in accordance with our current conception of the good life we will have an interest not only in individual free choice, but also in the existence of a specific and sufficiently stable cultural structure, for it provides us with the range of choices that render my free choice meaningful and valuable in the first place. Kymlicka conceptualizes such stable cultural structures as a secure *context of choice*, i.e. as a "secure cultural context from which individuals can make their

---

freedom and equality. For as I contended: people have *equal* rights to those (basic) conditions which they need in order to be *free* agents and as such can unfold their full humanity.

[53] See Kymlicka (1989, p. 162).
[54] *Ibid.*, p. 10.
[55] See Rawls (1971, pp. 395–452). For Rawls, "a person's good is determined by what is for him the most rational plan of life given reasonably favorable circumstances" (*ibid.*, p. 395).
[56] Kymlicka (1989, p. 165).

choices".[57] Some authors like Jeremy Waldron doubt Kymlicka's claim that there is such a thing as "cultural structures" let alone that such structures should be guaranteed so as to make sure people can make meaningful choices.[58] This scepticism of Waldron arguably misses the mark though. There can be no doubt that we are generally the product of the people and community around us. For the formation of our respective identities and for the conception of ourselves as *continuous* entities, we depend at least on *some* degree of continuity of the constellations around us.[59] That there *are* continuous (i.e. stable) structures of this sort around us and that these structures and its continuity are worthy of being (socially) protected is hard to contest, at least unless one is willing to identify life — counterfactually — with "isolated individual existence".[60]

The concept of a secure *context of choice* will be important when discussing the question of why and when exclusion is legitimate. But more interesting at this stage is its theoretical merit: the idea that cultures provide for a secure *context of choice* helps transform a vague concept like culture into a somewhat more tangible good — a good that the individual values and in which she has an *essential interest*. Through such understanding of a *secure context of choice* as a good the individual has an essential interest in, the significance of culture is finally brought home to, or translated into the language of, liberalism. Thus, a first answer to the question of why communities and their cultural structures are of value to the individual, and why they are thus meaningful to the moral approach pursued in this work, is this:

> Liberals should be concerned with the fate of cultural structures, not because they have some moral status of their own, but because it's only through having a rich and secure cultural structure that people can become aware, in a vivid way, of the options available to them, and intelligently examine their value.[61]

Departing from the way the *secure cultural context of choice* was just conceptualized, one may now consider the clarifying distinction between (1) "culture" as the *character of historical communities*, as opposed to (2) "culture" as a *context of choice* for individuals. With this distinction, Kymlicka brings to light two meanings and common usages of the term

---

[57] *Ibid.*, p. 169.
[58] See Waldron (1995, p. 106).
[59] See Sitter-Liver (2003, p. 39). See also Haker (2002, p. 394), on whom Sitter-Liver draws to make his forceful points on the relevance of communities for the constitution of personal identity and the self.
[60] Sitter-Liver (2003, p. 39).
[61] Kymlicka (1989, pp. 165–66).

"culture". As it will turn out, these two meanings will help draw a very clear line between, on the one hand, the way in which communitarians attribute significance and meaning to culture and, on the other hand, the way in which liberals should include culture in their theorizing.

Clearly, this distinction has important implications for matters of immigration. Where the focus is on the (1) *character* of a community, and if this specific character in its present fixed form is taken to be of (moral) value, then once norms, ways of life, etc. change (for example through immigration), the community and its culture may be said to have suffered a (morally relevant) loss. The change that invariably takes place through immigration by that view appears to be closely linked to the possibility of cultural loss, and it seems inevitable that immigration is here viewed as something negative or problematic at best. Michael Walzer's "communities of character" as discussed above provide a fitting example of this understanding of culture. When he claims that a community is almost invariably at risk through immigration, he may thus be right to the extent that he draws on such an account of culture.

If, on the other hand, culture is valued as the (2) *context of individual choice* which provides us with meaningful options and the "ability to judge for ourselves the value of our life-plans",[62] then immigration can be seen in a more open-minded and relaxed way. Even where changes take place through the influx of immigrants and through the new cultural elements they bring along, the *context of choice* can normally remain in order. The possibility that the cultural structure is changed in its character is not a problem *per se*, it constitutes no loss. Rather, new views and ways of life — providing for a wider spectrum of options from which to choose — are added. The range of choices available to the individual is, potentially and *ceteris paribus*, enlarged. On that view, immigration is clearly not a problem in itself.

It does become problematic, however, once there is reason to assume that through the influx of immigrants — for example when entering too quickly, in too large numbers, and from culturally distant places — the *context of choice* could become destabilized. It becomes problematic at that point when the *context of choice* would become poorer rather than richer in terms of the options it offers to the members of the respective community. By itself, the very possibility of such negative developments presents a strong reason in favour of the view that states should in principle have the right to exclude.

---

[62] Kymlicka (1989, p. 166).

In sum, it has become clear by now that culture matters morally and that a rich *context of choice* is a good that is needed and valued by the individual. In principle, this *context of choice* can be enriched through immigration. However, it also turned out that the individual can find the richness of this context eroded, or rather at risk of eroding through certain forms of (rapid and large-scale) immigration. The possible costs of this erosion are far-reaching and concern nothing short of the functioning of communal and democratic life.

What fosters this erosion is, arguably, a reduction in the level of *social trust* among citizens that comes with certain forms of immigration. The psychological claim that I will draw on in the following section is that in response to certain forms of immigration people will tend to trust less in their fellow citizens because there is an increasing number of people within the community of whom they assume that they do not share certain (basic) cultural notions, let alone any other feature determinant of their own identity.[63] In line with the political approach as chosen in this work, I will not judge this psychological claim morally and I will not criticize individuals whose attitudes (toward immigrants) confirm this claim (as long as such attitudes to not rest on racist assumptions). Rather, one is well advised to assume its general empirical validity (which I do) and then assess in a practically oriented way what it means for the moral analysis of immigration policy. This is the task of the following section.

3.2.3. The Proper Role of Shared Identities and Social Trust

> The big question here, hotly disputed among political theorists, is whether citizenship alone is a sufficiently strong cement to hold together a democratic welfare state, whose successful working depends upon relatively high levels of interpersonal trust and co-operation, or whether it is also necessary for the citizens to share a national identity of the kind that common nationality provides.[64]

Notions of *national identity* play an important role in the ethics of immigration and we have already come across *exclusive* conceptions of such shared identities in the previous sections. The following inquiry into the meaning and proper place of *national identities* in normative reasoning on immigration will complement and substantiate the previous concept of a secure *context of choice*. I will point out to what extent the prevalence of a certain sense of "common identity" or "national

---

[63] See Miller (2016, p. 145). This psychological claim has already been empirically substantiated in the introduction.
[64] Miller (2008, pp. 371–390).

identity" is helpful (or even indispensable) for providing the substances of important moral rights. To the extent that national identities can erode where diversity increases too quickly through rapid and ample immigration, countries face a trade-off between paying consideration to the interests of migrants in entering the country on the one hand and, on the other hand, sufficient levels of *social trust* (as facilitated by shared identities) on which the functionality of relevant societal institutions depends. As Margaret Moore points out, with levels of social trust being sufficiently high, the social and democratic institutions of the state tend to function more smoothly, the political community is better able to solve collective action problems, and it is easier for a state to establish demanding institutions conducive to social justice.[65]

As a first approximation then, what may a national identity legitimately consist in? First of all, it must be marked off from what it is not: it is not that kind of identity that I above referred to as an *exclusive identity*. It does not build on the fixed *character* of a given cultural community at a particular and arbitrary moment in time. It cannot be adduced for the exclusion of some people simply because they are, for example, not white or not fluent in a given language, etc. It does not draw on a set of traditions, rituals, and practices from which all people are excluded who simply do not reflect those traits. It is not an illiberal identity where the nation figures as sacred and as an ultimate value. And finally, it is not promoted by state authorities in an aggressive and coercive fashion as in some Eastern European countries today.

So how, in positive terms, should a national identity be conceived then? First and foremost, any such identity must have at its core the recognition of and respect for the rights of all people, i.e. not just of those people who share the identity in question. Moreover, a national identity should (and clearly can) be inclusive: it should be open and pluralistic in nature, and it should leave room for minority groups who may practice any culture they like, provided such practices are not in direct contradiction to the liberal norms at the heart of the promoted national identity.[66] Any such practices must be compatible with the equal rights and freedoms of all others. Indeed, that the prevailing identity is an inclusive one is of vital importance especially in today's pluralistic societies in so far as part of such identity's *raison d'être* is

---

[65] See Moore (2001, pp. 74–101) and Miller (2016, p. 145).
[66] See Banting and Kymlicka (2004, p. 252).

precisely "to establish trust between groups who might otherwise be disposed to treat each other with hostility or disdain".[67]

The vital source of such inclusive identities are shared political values, unfixed cultural values, a common language that transports meanings and makes public discourse viable and lively, and the social institutions that operate in that language. On that account of identity, culture matters a great deal, but it is not fixed and not exclusive: the value of culture is determined by its ability to contribute to a rich *context of choice* and thus to the effective fulfilment of human rights and freedoms within the community. The common identity as championed here must broaden the range of opportunities, worldviews, and mindsets available for free individuals rather than narrow down the valuable choices they have and want. Moreover and very importantly, the individual members of community who are subject to the prevailing identity must enjoy the human right of free speech, because only this right allows them to question the actual value of the current national identity. Free speech makes sure that any national identity may be constantly challenged, that it may be criticized as too exclusive and that it can be altered through public debate. In countries where inclusive national identities are endorsed by a government there must be enough room for minority cultures.[68]

There is an important and frequently raised accusation against national identities: namely that all forms of national identities, whether inclusive or exclusive, tend to be artificial constructs. What is certainly true is that such identities are, in relevant respects, forged through politically steered "nation-building" projects in the course of which a single language and cultural ideas are extended over the state's

---

[67] Miller (2016, p. 145).
[68] See Kymlicka (2001, pp. 258–59). Note finally that even a *common history* may form part of such an inclusive national identity. But to avoid that an emphasis on history leads to exclusivity, it is crucial to understand (or frame) a shared history not in terms of the biological ties that present generations have with past generations. Rather, present members—no matter whether their ancestors were members of the historical community in question—may share and identify with a common history in the sense that for their current freedom and well-being they depend on the community's historic achievements (at least in liberal democratic countries): they benefit for example from a solid constitution, from the values at that constitution's base, and from (stable) social, political, and economic institutions. Current members may and should appreciate these achievements as products of the community's history, and as sources of their own individual freedom. Such achievements, where they exist, can be a legitimate integrating factor.

territory and the people residing on it.[69] And there can be no doubt that they are at least partly constructed, imagined, and to some extent based on myths.[70] But this myth-based character is not necessarily problematic. It doesn't matter if the narratives behind a national identity are not or are only partly accurate as to historical or cultural facts. What matters here is that the narrative creates shared meanings—or identities—which then have important functions for the successful working of society.[71] The central consideration then shouldn't be whether "national identities embody elements of myth, but whether they perform such valuable functions that our attitude, as philosophers, should be one of acquiescence, if not of positive endorsement".[72] We should endorse those identities that are conducive to a social environment in which certain norms and attitudes needed for the effectuation of important rights can thrive, like mutual respect and the willingness to cooperate and share.

Now what exactly are the implications of those previous assessments for the morality of immigration? To be sure, whether an exclusive national identity or an inclusive national identity is adopted has decisive implications for immigration policies. Where *exclusive* understandings of national identity prevail, as in some Eastern European countries, the national identity will serve as the justificatory ground for excluding outsiders who do not portray a certain set of characteristics (culture, religion, fluent and authentic command of language, ethnicity) generally believed to be integral to the national identity. An immigration policy grounded in the aim of preserving such an account of (exclusive) identity will be unacceptably restrictive. The rigidity and arbitrariness that characterizes such border policies based on an illiberal conception of national identity must be rejected as inconsistent with the here upheld assumption of people's equal rights. This is so because some people will never be given even the chance that due consideration is given to their claims for admission, let alone that they get a comprehensible justification for their exclusion. This is intolerable from the standpoint of equal human rights postulated in this work.

If, on the other hand, common identities are *inclusive*, the tendency that immigrants make the country more heterogeneous is not a problem *per se*. And yet, immigrants can in principle be excluded when they do not accept the norms and values at the base of the common inclusive identity. Exclusion is an indispensable authority in so far as the *context*

---

[69] Kymlicka (2001, p. 255).
[70] See Anderson (1983).
[71] See Miller (1992, p. 92) and Miller (1995, pp. 35–47).
[72] Miller (2000, p. 31).

of choice can come at risk through immigration. What existing members potentially have at stake through immigration is an essential good, and if current citizens have in principle a right to protect such an essential good, which they presumably do, they must also have a right to exclude.

One further remark must be added here, and it addresses all those who regard even the inclusive form of common identity as standing in the way of more encompassing, global identities that could give rise to stronger feelings of regard for, and empathy with, fellow human beings in other parts of the world. Such higher-level identity might allow for more generous and effective international redistribution to help the global poor, an outlook many moral philosophers are likely to endorse.[73] So why not endorse it? As a matter of empirical fact, there is simply not enough of such international identity at the moment. It is, moreover, difficult to build it. And finally, there is after all no reason to believe that identities within the state stand in any principled opposition to more broadly conceived internationalist and cosmopolitan identities. Rather, inclusive national identities may be seen as the natural starting point for such more far-reaching empathy. This presents a further reason why one may value common identities at home and the empathy (and equity) they make possible. Of course, societies should make an effort of extending these feelings to ever greater groups of people. But as long as an international identity strong enough to support ambitious schemes of global redistribution doesn't exist, and is indeed not foreseeable, one should be wary of those who warn even of the here proposed inclusive national identities: "From the perspective of cooperation between people, nations are not selfish impediments to global citizenship; they are virtually our only systems for providing public goods."[74]

3.3. Special Duties and the Right to Protect: On the Function of Social Trust in a System of Moral Division of Labour (MDL)

In the previous section I made no attempt at integrating my findings on identity and social trust into this work's broader normative framework, and I so far indicated only tacitly what follows normatively from the observation that national identity and the ensuing *social trust* among citizens can potentially come at risk through certain forms of immigration. So how can the previous section's insights be lifted to a more explicitly normative level?

---

[73] See Collier (2013, pp. 235-36).
[74] *Ibid.*

Let me take a step back first and consider the work of American philosopher Robert E. Goodin. Similar to my conception of *states* as dividing moral labour among them (MDL), Goodin also takes individual states as the appropriate units to which special responsibilities should be distributed. National boundaries figure in his account as "useful devices for 'matching' one person to one protector".[75] In this way, with every person being assigned to a particular state, Goodin expects to achieve "maximal fulfillment of everyone's general duties toward everyone else worldwide".[76] Surprisingly though, he does not offer an account of what it is that makes this expectation plausible. So what is the *unspoken assumption* that makes plausible the view that nation-states are appropriate bearers of special responsibility? What is it about states and their societies that makes them "useful devices" in the first place?

Here, the above contentions on national identity and social trust come back in. As I shall contend, any system under which moral duties are mediated will depend for its success on its ability to sufficiently *motivate* people to accept the special responsibilities they are assigned. So when Goodin speaks of national boundaries as "useful devices" for matching right-holders to their protectors, he neglects what vital resources there need be in place within these national boundaries to make such boundaries useful devices in the first place. One vital resource is arguably a common identity that ties citizens together and that allows for social trust, mutual regard, and solidarity among citizens. A certain level of social trust among the citizenry is necessary to "provide public goods that would otherwise not be well supplied by a purely market process".[77] This underlines that rights depend on certain forms of community.

Surprisingly, such considerations on the important role of common identities and trust have no systematic place in most arguments on moral division of labour.[78] This is clearly a shortcoming. For after all, the social trust view can explain *why* individual nation-states are effective devices for securing human rights: it is in political associations tied together by common identities and social trust that people are sufficiently motivated to fulfil the assigned *positive* duties toward their fellow citizens. Individuals are much more likely to fulfil the obliga-

---

[75] Goodin (2008, p. 275).
[76] *Ibid.*, p. 274.
[77] Collier (2013, p. 63).
[78] Although centrally concerned with moral division of labour, authors like Goodin (2008), Shue (1988), or Weitner (2014) do not integrate an account of (common) identity and social trust systematically into their arguments.

tions assigned to them where sufficient levels of social trust exist in society.

If one assumed a world in which all people acted in a perfectly moral way, in the sense that people felt the same sense of solidarity toward all other people in the world, then there would be no insufficient motivational capacities standing in the way of assisting even distant strangers and of improving the global institutional order in such a way that poverty and misery could be fought sustainably. Our present world is not like that though. It is not ideal, and it is in such a non-ideal world that one should take the existing empirical understanding of the importance of social trust very seriously.

The MDL-model is then best understood as a way of dealing with a world that is non-ideal. It calls for the existence of a system of states in which groups of individuals, sharing conceptions of the good that bind people together, take care of each other. For in more global arrangements people are presumably at least less likely to take care of each other. In this sense, MDL is a *response* to a non-ideal world. Yet, some authors seem to understand it in just the reverse way. They construe of a system of individual states as being, only *under ideal conditions*, the best instrument for the aim of securing human rights (globally).[79] As Goodin contends with a view to the imperfect world of today, "it is often — perhaps ordinarily — wrong to give priority to the claims of our compatriots."[80] Arguably, this is to miss the very point of why to draw on MDL as a theoretical construct. To make normative sense of a world in which people will generally prefer to primarily help those fellow citizens whom they trust is a concession to reality, not an assumption to be abandoned once the world shows its ugly face. The importance of social trust should enter the account explicitly, as a *sociological presumption*. To divide moral labour between sub-units where people have sufficient trust in each other is a way of dealing with the fact that people have not yet developed sufficiently cosmopolitan identities. It is thus under *realistic* conditions that individual nation-states figure as the most feasible and effective devices for achieving full coverage of human rights.[81]

Let me sum up the argument so far. It comes in three steps. (1) If my assumptions on MDL are correct in so far as the *assignment* of special duties within the context of individual states is a practically effective way of *securing the human rights* of all people, and (2) if my assumptions

---

[79] See Weitner (2014, p. 154).
[80] Goodin (2008, p. 275).
[81] Note how this reasoning resonates with the *political approach* adhered to in this work.

on social trust (and by the same token on common identities) are correct in so far as citizens' *acceptance* of such special duties hinges for its realization on the motivational resource of social trust among citizens, then (3) it is a concern for human rights that at least on a *prima facie* basis warrants the attempt to foster and *protect* social trust among citizens. This reasoning leads to the formulation of an immigration-related *right to protect*: to the extent that immigration restrictions are sometimes necessary to protect social trust among citizens, such restrictions have a *prima facie* moral justification that draws its strength from a concern with human rights and its effective realization.[82]

It seems that both in the ethics of immigration and in public debates, calls for open borders tend to neglect the way in which immigration can reduce social trust and thusly social justice within societies. Arguably, one cannot (realistically) argue both for open borders and for a system of MDL where human rights are effectively secured. Open borders would *undermine* the rationale of MDL. To the extent that open borders compromise individual states' capacity to redistribute and protect the most vulnerable, open immigration seems to be detrimental rather than conducive to the realization of core human rights.

Let me finally explain why a risk-ethical perspective should inform any concern with immigration and social trust, or more precisely with the relationship of these two. The justification for a risk-ethical assessment is that there remain important uncertainties regarding that relationship between immigration and social trust (or mutual regard). As Collier points out, "beyond some *unknowable* [my emphasis, J.K] point the losses from reduced mutual regard are liable to increase sharply as thresholds are crossed at which cooperation becomes unstable."[83] Indeed, once unstable, there is a point where further immigration does not lead to more instability, but to breakdown: "Cooperation games are fragile because if pushed too far they collapse. In fancier language, equilibrium is only locally stable."[84] This suggests that at some point the cost of admitting a certain number of further migrants, of allowing the composition of members to become more diverse, can

---

[82] Recall once more that the here formulated right to protect is contingent on the circumstance that the level of social trust indeed decreases with the influx of a given number of immigrants. However, the right to *control* as I will defend it in the following section is not contingent on empirical circumstance. After all, the general right to exclude draws its plausibility both from the elaborations that lead to the right to protect *and* from the elaborations that lead to the right to control.
[83] Collier (2013, p. 63).
[84] *Ibid.*

be substantial: where this tipping point is reached, the freedom of citizens could be reduced dramatically and potentially irreversibly through the admission of further (non-refugee) migrants. Maybe Collier goes too far when he concludes from empirical evidence "that continually increasing diversity could at some point put [...] critical achievements of modern societies at risk".[85] Yet, his assessment underlines that much is at stake, it grounds the need for a risk-ethical perspective on immigration policy, and it corroborates the assessment that legitimate states should have the authority to exclude.

3.4. Special Duties and the Right to Control: Ryan Pevnick's Associative Ownership View

> Like the family farm, the construction of state institutions is a historical project that extends across generations and into which individuals are born. [...] The citizenry raises resources through taxation and invests those resources in valuable public goods: basic infrastructure, defense, the establishment and maintenance of an effective market, a system of education, and the like. As with the system of irrigation on John's farm, these are goods that only exist as a result of the labor and investment of community members.[86]

By creating and contributing to the (public) institutions of their state (from the education, health, and social system to the core political institutions), the citizens of a political association find themselves in a position to make credible *ownership claims to* these institutions and to the room of justice that is given shape through these institutions. The citizens are entitled to make ownership claims to such institutions and to the goods they provide for because without the citizens' shared efforts they would not exist. This rather commonsensical notion is encapsulated in the above quoted passage of Ryan Pevnick. I shall now present what Pevnick has introduced into the morally-normative debate on immigration policy as the *associative ownership view*. I will adopt and draw on this *associative ownership* view as a persuasive way for justifying a state's right to self-determination and therewith to exclusion. It will inform my case for a *right to control* immigration and it will in this way corroborate not only my earlier elaborations on why political associations are entitled to self-determination but it will also complement my overall contention that states have a right to exclude. At the end of this section, I will point out how the *associative ownership* view and the account of self-determination fit together with the previously proposed (normative) model of states as dividing moral labour

---

[85] *Ibid.*, p. 244.
[86] Pevnick (2011, p. 40).

among them (MDL). Indeed, the associative ownership view is meant to *complement* previous approximations to the idea of self-determination (e.g. my elaborations on the idea of a "room of justice"). But other than those previous approximations, the following account will be applied more straight-forwardly to the issue of immigration control.

Pevnick begins his account of the *associative ownership* view with the assessment that it is rather intuitive to hold any association of people who collectively created a good to be in a position to claim ownership over that accomplished good. (And indeed, one might ask: who would be in a better position to claim ownership than them?) Pevnick takes as an example such *voluntary* congregations as a youth education programme and a widows support group. It is only because of its members' shared effort that these programmes have become a reality.[87] To bring them into existence they spent physical and intellectual effort, financial resources, and their time. They added their labour to the programme (i.e. a good), while others did not, and this makes them rather than any other persons the natural candidate for making a credible ownership claim. At the heart of this claim lies what John Locke observed with a view to individual persons: "The *labour* that was mine, removing them out of that common state they were in, hath *fixed* my *Property* in them."[88] No uninvolved person or group of people could make as powerful a claim to ownership and thus to *control* the future course of the established programme or institution. As owners, they can make a *special claim* to determining the future shape of such institutions, and part of this self-determination is presumably to decide on the association's *membership*, which implies the right to exclude non-members.[89] Such exclusion would, at least in principle, not conflict with the basic assumption that all people are morally equal and that they have equal moral rights. Moreover, such ownership-based exclusion is not subject to the kind of criticism that is rightly raised against the way in which nationalists in, for example, Eastern European countries lay claim to the territory of a given country. In such cases, the ownership

---

[87] See Pevnick (2011, p. 33).
[88] Locke (1988, p. 280).
[89] Here is a helpful analogy that Pevnick offers in order to further elucidate the reasonableness of his *associative ownership* view. The assumption his analogy draws on is that I, or any other person to whom it may apply, did not contribute relevantly to the achievements of Google. In my own case, that is indeed a rather uncontestable fact. Therefore, "the founders of Google do not fail to treat me as an equal when they, in the process of refusing my claim to their bounty, point [that fact] out" (Pevnick, 2011, p. 36). Google has a right to exclude me, and so do political associations to whose institutions and achievements I have not contributed.

claim is based on the mere "belief that this part of the world is the homeland of the majority national group".[90] This way of making a claim to ownership is indeed hardly convincing. It is unacceptable for the same reasons that lead me to reject *exclusive* conceptions of national identity and their illiberal (and potentially racist) implications for immigration policy. There can be no doubt that Pevnick's ownership-based reasoning goes into a very different direction.

As a confirmation of the modest character of Pevnick's argument, it is important to note that Pevnick's case for exclusion is normatively limited, at least in its application to political associations like states. For as Pevnick makes clear, any entitlement to exclusion on the basis of the *associative ownership view* can only be a *prima facie* one: there may be situations where other considerations (concerning human rights, for example) could limit the normative thrust of the property-related considerations in question. So as a qualification to the *associative ownership* view Pevnick notes:

> [R]eason has *not* been provided to claim that laboring on something or bringing it into being gives one full entitlement to it. There may well be all kinds of limitations on such ownership claims. For example, we might insist that those excluded are not harmed by this exclusion, that the needs of others impose limits, and perhaps there are other kinds of limitations as well. Thus, the associative ownership argument justifies members' *limited* claim of ownership over the congregation and its institutions.[91]

However, as long as there are no competing special claims and if in the process of creating the goods no other people were harmed, there is good reason to see the political association in a "special ownership relationship"[92] with the institution they brought into existence. This special ownership relationship grounds the political association's entitlement to self-determination over the established good.[93] But does this logic hold for *states*? After all, states are—other than the voluntary congregations in Pevnick's example—*non-voluntary* and *intergenerational* political associations.

Regarding the first question, Pevnick contends that an application of the *associative ownership* view to the case of states is possible. In order to make that contention, he draws on the example of Mary, who, unwilling to pay for the high costs of university education, decides to kidnap a group of promising political theorists. She forces her captives

---

[90] Kymlicka (2001, p. 257).
[91] Pevnick (2011, p. 35).
[92] *Ibid.*, p. 36.
[93] See *ibid.*, p. 34.

— involuntarily associated as they are — to work together on lectures on political theory. The theorists are finally freed, and the lectures they collectively conceived during their time in custody became a commercial success. Now, despite the fact that they did not enter the joint work voluntarily, Pevnick contends, the political theorists own the results of their collective effort. The labour they put into the lectures gives them an ownership right to them, which also entitles them "to the profits that result from the sale of the lectures as well as future rights to, and control over, the (intellectual) material therein. This suggests that lasting claims of ownership can originate in non-voluntary associations".[94]

Regarding the second question, Pevnick affirms that the *associative ownership* does hold in the face of the circumstance that states are *intergenerational* schemes of cooperation. To bring home that claim Pevnick again makes use of an analogy. He depicts an old-established family farm, of which little John becomes a member right at his birth, and to the privileges of which he can claim full and equal rights once he reaches a certain age. Pevnick contends that "John's family has a claim of ownership over the farm, one element of which is a right to make decisions regarding its future (including decisions about *who* will make such decision in the future)."[95] This justifies current members in passing membership rights; it justifies them in passing the right to control to John. It would be illegitimate for outsiders to approach John with the request to use, for example, a different irrigation system. One could object that it was pure luck that John now owns the farm. And yet, his ownership is legitimate because previous owners passed these special rights down to him. This grounds at least a strong presumption in favour of John (a *prima facie* right to control, if you like).

Now, turning to the case of immigration control, the *associative ownership* supports the case that states have a *prima facie* right to exclude outsiders. For as I see it, this view can serve as a further normative source of *special duties* toward fellow citizens (with the rationale behind MDL being another such source). It does so only indirectly though, because what *associative ownership* grounds is primarily a right to self-determination over the benefits of one's efforts and achievements, which is, admittedly, not directly and not necessarily a ground for assuming special duties. But in drawing on that right to self-determination, a political association is clearly justified in prioritizing members, if only it decides to do so. Being self-

---

[94] Ibid., p. 37.
[95] Ibid.

determined, citizens can simply decide to owe more to one another than to non-members. This may, *in effect*, be the same as saying that they have *special duties* toward each other.

Note finally that Pevnick's *associative ownership* view can be brought in line with my above contention that individual states ought to take responsibility for the protection of the human rights of an assigned group of people (MDL), namely their own citizens.

Recall that MDL is motivated by the rationale that the universal duty to protect human rights is most effectively fulfilled when it is assigned to agents that are in a promising position to actually fulfil (an assigned share of) it. Now, the normative notion (and the *practised* standard) that individual societies have a *prima facie* entitlement to its proceeds is indeed a central aspect of the *incentive-structure* that makes MDL so effective in the first place. In this way, *associative ownership* can be readily integrated into MDL. More still, MDL and *associative ownership* can lend themselves mutual support. While under MDL states' right to self-determination is granted *so that* states can effectively fulfil their task of building effective institutions for the protection of human rights, the *associative ownership* view moves the other way around from *is* (fact, lat. *facere*) to *should* (normativity): citizens first build or maintain effective institutions as elements of their room of justice so that they can then make a forceful claim to owning (i.e. controlling) them, and thus to excluding others from them.

3.5. On the Special Claims of Migrants

There may be cases in which the claims of *particular* outsiders to admission could outweigh the claims of insiders to exclusion. How are such cases characterized? They are marked by the important difference that the migrant in question can make a *special claim* toward the political community.

To characterize what makes an outsider's claim a "special" one, it bears taking one step back. Non-refugee migrants were in this account conceptualized as people who do *not* migrate because they were forced to do so or because they seek to secure fundamental (or essential) needs and interests. In other words they are neither war refugees nor "economic refugees". Migrants as here defined left their country *"voluntarily"*, *i.e. without basic interests at stake*. This may in principle hold true for migrants with a *special* claim (as here understood) as well. As I will contend, what distinguishes them from normal migrants though is that they stand in a *relevantly disadvantageous relationship to the destination country*.[96] This is their special claim. Potentially, such a special claim could outweigh the potential host country's (good) reasons for exclusion and could thusly compromise the legitimacy of a state's exertion of its right to exclude. This is a decisive point, for what the aforesaid suggests is that there is, after all, a sense in which a state would do wrong by excluding certain particular migrants.

Before I turn to the implications of this assessment, I will dwell a bit more on the character of a "relevantly disadvantageous relationship", which — if uncompensated for — grounds a special claim which in turn could outweigh the *prima facie* right to exclude as regards the particular case. The basic assumption is intuitive: provided there is a relationship between the particular migrant and the destination country which is evidently to the detriment of the migrant and which is related to her migration then the existence of such a detrimental relationship would lend her claim toward the destination country extra normative weight. The source of this normative weight is corrective justice (CJ) as introduced earlier. A concrete example of such a special claim would be a migrant's assertion that her poverty, which clearly influenced her decision to leave her former country, was brought about by, or persists among, other factors due to international economic institutions (or rather: structures) which the destination country plays a relevant role in shaping. If the destination country's actions or inaction (for example in negotiating the terms of international trade) harm the migrant, then, it seems, the destination country owes some form of compensation or remedy to the migrant. Hence, the assumption is that where a country of destination was involved in causing the applicant to migrate in the first place, the destination country cannot simply exclude that person on the very same argumentative basis on which the exclusion of all other, non-related migrants could be justified. It is indeed a moderate claim to say that once a country is confronted with the special claims of particular migrants the rationale behind exclusion must at least minimally take account of that relevant circumstance and cannot simply hold on to those approved justification strategies applied to normal migrants who cannot make such a special claim. My proposal is this: if there is indeed a detrimental relationship grounding a special claim (and I ignore here all the difficulties related with the attempt of reaching consensus on what would constitute such a detrimental relationship in the real world) and if the country wants to retain the possibility

---

[96] This notion of a special claim resonates with my characterization of a special right as proposed in the second chapter. By drawing on Shue, I there defined special rights as rights that can be claimed "because of an act, event, or relationship of which a causal or historical account can be given" (Shue, 1988, p. 88).

of legitimate exclusion even in the face of such a detrimental relationship, then the country must have made good, i.e. appropriately compensated for, the disadvantages that it brought about. This assessment follows from a simple application of a corrective justice perspective.

However, no migrant will be automatically entitled to a right to immigrate simply by pointing to present day inequalities that characterize the world economy. Rather, whatever responsibility toward migrants results from a presumed unfairness in global economic structures, all that would follow from such an observation is that countries which benefit from and shaped existent global economic arrangements owe *something* to the migrants, but this something is not necessarily the right to immigration. Rather, to allow the migrant to immigrate is only one possible way of compensating, i.e. of discharging whatever that person is owed.[97] From the perspective of the owing country there are in general at least two ways of fulfilling a presumed compensatory duty toward the migrant. Either (a) it discharges the duty of compensation in some non-immigration-related form, say through assistance in situ, or (b) it discharges the duty by allowing the migrant to immigrate. The political community's assumed duty toward the migrant is thus a *disjunctive* one. What this disjunctive form further suggests is that if the country does not fulfil its presumed compensatory duty by way of (a), it would have to let the migrant immigrate (b). And if it does not discharge it in some non-immigration-related way (a) and at the same time does not discharge its duty by admitting the migrants to the political community (b), then the country hinders the migrant from getting the compensation that she has a right to. This would constitute a moral wrong, pervasive though it may be in the present world.

One may also frame the just conceptualized disjunctive duty in terms of primary and secondary duties. Such a division into primary and secondary duties is frequently found in current ethical debates on immigration.[98] The logic is that if wealthy countries are involved in detrimental global processes or structures (from which they benefit) and if they do not fulfil their primary duty of rendering such structures less disadvantageous and of improving the situation of the global poor

---

[97] In much the same way, there are in principle various ways of providing a refugee with what we owe her, but in practice, in the immediate situation of her arrival, the range of options by which to help her is narrow, and most often probably limited to giving her at least temporary shelter.

[98] For perspectives in which this distinction structures the argument on immigration policy, see Bader (2002), Goodin (1992), Pogge (2002), and Schlothfeld (2002).

in situ, then they are obliged to take in more needy people from those countries—as a secondary duty.[99]

> The goal of such exercises is precisely to put rich countries on the spot. The aim is to argue that, if arguments for international distributive justice are valid and if rich countries do not want to give generously of their money to meet the demand that those arguments impose, then they are morally obliged to pay instead in a currency that they hold even dearer.[100]

The practical helpfulness of this reasoning may remain questionable. But the perspective of primary and secondary duties that lies behind it leaves us with a useful lens for judging the legitimacy of immigration policies. And to the extent that the primary duty boils down to nothing short of the negative duty not to act against other people's basic rights to freedom, security, and well-being, it is hard to deny the general reasonableness of the view that a state's disrespect for the described primary duties must indeed result for that state in the compromised legitimacy of exerting its right to exclude. By understanding the duties toward migrants as disjunctive ones and by introducing the distinction between primary and secondary duties, one can see clearer how the (special) claims of outsiders enter the overall assessment. It should be evident by now why and in what sense the right to exclude is invariably only a *prima facie* one: namely that the legitimacy of *exerting* it can be constrained by the particular claims that particular migrants can press against the potential host state.[101]

---

[99] See Schlothfeldt (2002, pp. 93-95).
[100] Goodin (1992, p. 8).
[101] What this suggests is that the democratic decision of the host state on how to *exert* the right to exclude can be subject to moral criticism, for example when—as just pointed out—the excluded migrants hold a *special claim* against the host state. It may be judged that it was morally wrong to *exert* the right with a view to a particular migrant. And yet, such a critique does not reach to the verdict that the state *has* the right to exclude (at least as long as it otherwise remains a minimally legitimate state). So, for a state to exclude particular migrants can be wrong and acceptable at the same time. The right to exclude is one that the citizens of a given state *have* even when in a particular case its *exertion* contravenes their own particular duties. As political philosopher Oliviero Angeli observes in a similar vein, "exercising a 'right' is not identical with doing the morally right thing to do" (Angeli, 2015, p. 108). It may be morally wrong that Donald Trump, as president of the United States, pays so little attention to the truth value of his assertions, but few would doubt he nonetheless has a right to make such doubtful assertions, presumably because most people tend to regard the right to free speech as more important morally than the right not to be fooled. It appears then that "to have a right" is different from "having *good reasons* for exerting that right in this or that way" (see Angeli, 2011, p. 182). A

After all, this implies that provided the primary duties *are* fulfilled, no special claims can be brought forward against the state. Its right to exclude can then be exerted legitimately, quite simply because the excluding state acted in accordance with the equal rights of the people affected by its actions and, more specifically, in accordance with the equal rights of the particular migrant asking for entrance. In this sense, excluding the particular migrant would not extend a prior injustice against him (as it would be the case if primary duties were not fulfilled) and thusly would not be inconsistent with his equal rights.

In the next chapter, the concern will be with a group of migrants who make a very urgent claim to admission, namely refugees. What they have at stake are the substances of basic rights. Their exclusion, it seems, would be a most dubious thing to do from a moral standpoint. But by itself, this verdict offers little practical guidance. It could even turn out to be simplistic when confronted with an increasingly challenging reality of global refugee flows. What follows is an assessment of our moral obligations toward that vulnerable group of refugees. My previous elaborations on the right to exclude will inform that assessment in important respects.

---

right is valuable in so far as it "opens up and protects a range of options to choose from, including bad ones" (Angeli, 2015, p. 108). The right to exclude that I argued for in this chapter is essential in precisely this sense. It allows the individual members of community to protect a room of justice in which each can enjoy several other rights and a degree of individual autonomy that they could otherwise not enjoy. The right to exclude must be a far-reaching authority, for what it protects is essential.

Four
The Special Case of Refugees

> [M]ost theorists of closed borders admit that there are circumstances under which the general right to exclude outsiders would become unreasonable to exercise.[1]

> Whatever principles or approaches we propose, we should always ask ourselves at some point, "What would this have meant if we had applied it to Jews fleeing Hitler?" And no answer will be acceptable if, when applied to the past, it would lead to the conclusion that it was justifiable to deny safe haven to Jews trying to escape the Nazis. This approach will not settle every question about refugees that we have to consider, but it will give us a minimum standard, one fixed point on our moral compass.[2]

In this chapter, the moral questions surrounding the issue of refugees will be approached systematically. There are several interrelated issues I will be concerned with. First, what is morally special about refugees and why is their claim to admission a particularly strong one (see 4.1)? Based on the results obtained from addressing this first question, the second concern is *what we owe to refugees*. What kinds of moral duties do we have — or, depending on concrete contexts, could we have — toward them (see 4.2)? And how should the obligations toward refugees be distributed among states (4.3)? Finally, the third concern is with the *scope* of any such duties toward refugees (see 4.4). Are there limits to what we owe to them? And if so, how can they be specified?

4.1. What is Morally Special about Refugees?

Refugees make a very strong claim to admission. When discussing the possibility of legitimate exclusion in the previous chapter, the individuals I had in mind were non-refugee migrants. Such previous discussions centred, for example, on economic migrants, who were characterized as moving rather voluntarily or — as concerns migrants who leave their country to escape situations of poverty (which of

---

[1] Blake (2005, p. 232).
[2] Carens (2013, p. 194).

course may greatly vary in degree) — as at least *less forced* or compelled to leave their home country than refugees are. It is important to note that, in principle, the reasons that ground a *right to exclude* hold just as well in the case of refugees. But as Michael Blake suggests in the opening quote to this chapter, it will sometimes become questionable for states to *exert* that right to exclude in just the same way they would (or could) exert it in the case of non-refugee migrants. Being forced to move and with their basic rights at stake, refugees make a human rights-related claim to admission that urges us not to *exert* the right to exclude in an inconsiderate manner. Such exclusion would often turn out to be reckless and inconsistent with the bedrock assumption of refugees' equal basic rights.

Refugees are by definition understood as being forced to move (which is not to say that the affected individuals lose all sense of agency). Recall here Matthew Gibney's refugee definition which I introduced and adopted in the introduction and which is broader than the Convention definition:

> [Refugees are] people in need of a new state of residence, either temporarily or permanently, because if forced to return home or remain where they are they would — as a result of either the brutality or inadequacy of their state — be persecuted or seriously jeopardise their physical security or vital subsistence needs.[3]

Now, on the face of it, the elements included in this definition may already make clear enough what it is that makes refugees special. They cannot stay in the state where they are or, when they already left it, cannot return to that state because it is *inadequate* in the sense that it is incapable of or unwilling to provide for their security and subsistence. Much in line with this assessment, Andrew Shacknove conceives of a refugee as an individual who "inhabits a wilderness of acute deprivation where life is jeopardized by an extreme threat to minimum security or subsistence".[4] Indeed, from a more explicitly moral perspective the crucial point is that in their states of nationality they find seriously and permanently jeopardized those things which they have *basic rights* to, namely security and subsistence. Not to have one's basic rights adequately secured in one's home state comes down to simply not enjoying such basic rights in that state, even where the state honestly embraces a formal catalogue of human rights, but remains incapable of enforcing it. As Henry Shue reminds us, the "proclamation

---

[3] Gibney (2004, p. 7).
[4] Shacknove (1988, p. 136).

of a right is not the fulfillment of a right, any more than an airplane schedule is a flight."[5]

So once more, what is the strong moral claim that refugees bring forward? They can claim that in their home states they effectively do not enjoy basic rights and that they were compelled to leave that state for a new one in order to find those basic human rights secured in the first place. It is in this sense that refugees are in dire need of a new state. Only once they find their basic rights to security and subsistence protected in a new state can they enjoy any other rights.[6] In such situations where people are effectively deprived of their rights, they cannot make unforced choices. It is in this sense that one may speak of people who try to escape situations of extreme vulnerability as *forced* to leave. The condition of refugeehood is marked by an intolerable lack of freedom.

This lack of freedom, the deprivation of basic rights, and the resulting forced character of their migration arguably render the claim they make for admission a particularly strong one. But here it is important to make some decisive qualifications concerning the sense in which refugees can be said to be deprived of basic rights. For a person to be considered a refugee (in moral terms), it is clearly not sufficient to point to a temporary loss of certain substances of basic rights. Arguably, what grounds the refugee's strong claim to admission is not that she is deprived of this or that substance of basic rights in a *particular* situation but that those basic rights are permanently insecure or seriously threatened in the long run. In order to support that decisive claim, I will take a step back and draw on Alan Gewirth's distinction between occurrent freedom and dispositional or longer-range freedom as introduced earlier.[7] To lose one's occurrent freedom means that in a particular situation one cannot control one's behaviour by one's unforced choices. Now, as an important observation, there may in all countries be situations—or periods—in which people see their occurrent freedom interfered with or at least compromised. Think

---

[5] Shue (1996, p. 15).
[6] See Shue (1996, p. 30).
[7] Recall that the loss of longer-range *dispositional freedom* does not necessarily imply the loss of *occurrent freedom*. You will suffer a loss of *dispositional freedom* when you learn that your island will be submerged in the near future and when you have no other place to go to (for then you can obviously not make longer-range life plans), but you will still be free in the occurrent sense then. You are still free to take a walk on that island in the particular case, or to take a few pictures of that island, as a reminder for posterity of the island there once was.

of such catastrophies as Hurricane Katrina hitting the US in 2005 and leaving many people displaced and without access to basic needs. This absence of occurrent freedom, however, clearly did *not* qualify affected American citizens as refugees (by the above definition) because they (or at least most of them) retained what I refer to here as dispositional freedom: the *longer-range* ability to control the course of one's life. In the example of those affected by Hurricane Katrina, this longer-range freedom is expressed in the reasonable prospect of *regaining* in their home country the occurrent freedom they had temporarily lost.

Now think of situations where there is no such prospect. Syrians who during the civil war are threatened by several conflicting parties, and who are in constant fear of the governments' barrel bombs, are not free in the sense of dispositional freedom. They lack the substance of a basic right, namely *security*, and there is no clear prospect of regaining that substance any time soon: neither their government, nor the international community will be able to (or willing to) protect them where they are. What they — depending on the concrete region of Syria they live in — will often be compelled to do in order to regain their dispositional freedom is to flee that insecure region. After all then, by recognizing that a loss of longer-range freedom (as refugees face it) does not preclude the prevalence of (some degree of) occurrent freedom, it becomes clear why it is not a contradiction to conceive of refugees as at the same time *forced* (or rather compelled) to leave their country *and* as retaining a clear sense of *agency*.

This moral conceptualization of refugees may be *complemented* by introducing the notion of *standard threats*. As just pointed out by drawing on the distinction between occurrent and longer-range dispositional freedom, many refugees' deeper problem is not that they find themselves deprived of the substance of basic rights in a particular situation, but that they are not adequately protected against those deprivations from either occurring *again* or even becoming the new normal. In other words, they are often not protected against standard threats, i.e. against those threats "that could ordinarily be expected to prevent, or hinder to a major degree, the enjoyment of the initial right assumed".[8] Unless there are established certain social guarantees that protect people against such standard threats, one can hardly conceive of the people in question as enjoying the vital substances of basic rights *as* rights. Refugees lack such guarantees and they thusly lack effectively secured rights, and this leaves them vulnerable toward individuals, toward a brutal government and other state institutions, or toward destructive

---

[8] Shue (1996, p. 32).

natural influences. Normally, it would have been the responsibility of the state to reduce such vulnerability by establishing social guarantees, not "against ineradicable threats like eventual serious illness, accident, or death", nor "against every imaginable threat", but against *ordinary* and *predictable* threats.[9] Now, both people threatened by aggressive governments and people unprotected against standard threats, like certain environmental or climatic forces, can by this work's definition in principle qualify as refugees. What matters is *that* they are deprived of basic rights in the dispositional sense, regardless of the *source* of that deprivation and the *immediacy* of that deprivation.

Now, there are generally two possibilities why people may be eventually left without such protection against standard threats. Put simply, it is either (1) because of the state (which is *unwilling* to protect), or (2) because of the nature of the threat (in the face of which the state is *unable* to protect). Both unwillingness and inability constitute a form of *inadequacy*. Note how this observation fits naturally with the above refugee definition of Gibney, where the "inadequacy" of the refugee's source country is either (1) that it is *unwilling* to protect its citizens against standard threats (which may often mean that the state *is* the threat) or (2) that it is *unable* to protect its citizens against the particular threat. While possibility (1) is partly covered by the Geneva Convention, at least where a brutal state is the threat itself, possibility (2) is not. The fact that a materialization of either possibility would mean a deprivation of people's basic rights suggests that there ought to be remedies for both of them. Where international assistance or intervention is not undertaken or cannot remedy (1) and where international assistance and cooperation are not undertaken or cannot remedy (2), such people unprotected against standard threats will often see only one remaining remedy, namely to flee from that state, provided they have the financial means to do so and are not hindered from exiting that state.

4.2. What We Owe to Refugees

The contention in this sub-chapter is that we have a very strong duty to assist refugees, and I will present three reasons in support of this contention. These three reasons draw explicitly on this work's normative framework.

Before this analysis can begin, a few remarks need to be made on the possibility that, whatever duties we have toward refugees, such duties invariably come in the form of a disjunctive duty. As in the case

---

[9] *Ibid.*

of non-refugee migrants, there would then be either a duty to take the refugee in, or to help her in some other form that does her justice. It appears that there is in principle no reason to insist that our duties toward refugees can only be discharged by granting them asylum. As Christopher Wellman observes, "the presence of those desperately seeking political asylum renders those of us in just political communities duty bound either to grant asylum or to ensure that these refugees no longer need fear their domestic regimes."[10] And in much the same vein, David Miller notes:

> The lesson for other states, confronted with people whose lives are less than decent, is that they have a choice: they must either ensure that the basic rights of such people are protected in the places where they live — by aid, by intervention, or by some other means — or they must help them to move to other communities where their lives will be better.[11]

I agree that it is reasonable to conceive of our duties toward refugees as coming in a disjunctive form. In so far as in their home countries refugees are unprotected against standard threats and thus do not enjoy their rights in the sense of dispositional freedom, other states should help by granting shelter or by assisting them in situ. In either way, they would ideally regain their dispositional freedom. While in situ assistance may often be the preferable option, I will in the following sections concentrate on immigration-related duties toward refugees. To be sure, however, in situ assistance will eventually become a more central concern in subsequent chapters on climate migration.

4.2.1. Non-refoulement and the Right to Asylum

There is first and foremost a duty not to deprive refugees of the substances of basic rights, i.e. of basic security and subsistence. In a first step, I will discuss this first duty only with regard to those refugees who have "successfully" fled to a safe state and when there make an appearance as *asylum seekers*. In a second step, I will then ask whether that duty also has any practical relevance with regard to refugees who have not made it to our shores.

The duty not to deprive refugees of the substances of their basic rights may seem odd at first sight. Is it not the case that refugees are characterized as exactly those people who have *already* lost those substances? So how could we deprive them of these substances? Is the situation not rather such that we either help them, namely by providing them with those substances, or that we do not help them, in this way

---

[10] Wellman (2008, p. 129).
[11] Miller (2005, p. 198).

just leaving them as we found them (or as they found us), which would then mean that we hardly *deprive* them of anything by not helping them? These questions suggest an understanding of the practice of asylum policy that is rather unrealistic in most cases. It is simply not reflective of the reality an asylum seeker faces. Assume as an imagined (but arguably representative) example a group of refugees who, after making a most dangerous journey toward Europe, finally make their way to the territory of the European Union, say to a Greek Island. In the very moment they arrive they will in at least the *occurrent* sense be physically safe, they get shelter and no one will assault them. Further, one may presume that they will be provided with minimal subsistence, for it would be outrageous if any refugee starved on EU territory. Indeed, at the time of writing this text, such minimal protection was widely provided in the EU (by state and non-state actors). So taking this minimal standard for granted, it clearly *is* a meaningful question to ask what an application of the assumed first duty implies in the context of asylum seekers. It turns out we clearly *can* deprive them of the substances of their basic rights to security and subsistence. We would deprive them if we sent them back—back on the immediately dangerous route from which they came and/or back to the country from which they fled because their basic rights are not secured there. Toward those individual asylum seekers of whom we know they would risk life and limb in their home country, we thus seem to have a *negative duty* not to expose them to these threats by sending them back (and toward all asylum seekers we have a duty not to send them back on an immediately dangerous journey). In international refugee law, this duty is enshrined in the principle of *non-refoulement*.

I will now move beyond the principle of *non-refoulement* and make a more general case in favour of assisting asylum seekers. In order to argue for positive duties toward that group, I will draw on an analogy borrowed from Alan Gewirth. Assume Carr, an excellent swimmer, who is relaxing on a deserted beach. Close to where he is lying he has moored his boat to which a long rope is attached. While lying there as he does, he sees a swimmer some yards away struggling and in obvious danger of drowning. His name is Davis. Carr knows that there are viable options available to him for rescuing Davis. He could swim out to Davis, or he could use the motorboat and throw out the rope to Davis. Aware of the consequences, Carr takes neither of these options, and Davis drowns.

With a view to this pitiful ending, Gewirth comes to a verdict that is well in line with our intuitions: Carr should have come to Davis's help, and it is morally wrong that he did not. The precept supporting this verdict is that one should assist a needy person at risk of suffering basic harms[12] provided one is in a position to do so at no comparable cost and provided the bearer of the right (i.e. Davis) cannot help himself. The argument that Gewirth offers to justify this claim is a reflexive one whose core idea is captured in the following quote:

> Since every agent has a deep stake in his own well-being, he must hold that he has a right to it and that other persons also have this right. He must hence admit that he ought to refrain from interfering with their basic well-being, and where his inaction would interfere, that he ought positively to act to protect their well-being if he can do so without comparable cost to himself.[13]

Transferring the exemplary case of the two swimmers to the case of refugees, it is crucial to recognize that Carr (symbolizing the potential receiving state) finds himself in a *transaction* with Davis (symbolizing the asylum seeker), a transaction that is under Carr's control. Carr did not enter that situation voluntarily, but it is nonetheless a most relevant fact that he is now in it, and that he controls it. Carr is now in a position in which he will invariably *act on* Davis. For an agent to be considered in a transaction, "it is not necessary that he initiate the transaction, it is sufficient if he controls its course and thereby crucially affects what happens to the other person, who is hence his recipient."[14] This is an instructive conceptualization with a view to those situations in which a state is confronted with asylum seekers. They enter into a most immediate transaction with the potential host state, and that state cannot but accept this reality of a transaction.

In an illuminating passage that deserves quotation at length, Gewirth makes clear that to help the needy person—in our case to help the asylum seeker at our border (who, we may presume here, *is* objectively a refugee)—is no generosity, but a perfect duty with a clearly identifiable right-bearer and a clearly identifiable bearer of the correlative duty. Not to give to a needy person the object of her right which would alleviate that need is disrespectful of her rights, and would even amount to harming her:

> It is possible, however, to harm someone or interfere with his well-being by refusing him the help he needs in order to recover from an already bad situation. Such interference involves not that one turns an antecedent well-being into ill-being, but rather that one prevents the other person from attaining well-being through means that are under one's proximate control. One's refusal to help inflicts harm on the other

---

[12] By "basic harms" I mean setbacks to the substances of basic rights.
[13] Gewirth (1978, p. 228).
[14] *Ibid.*, p. 220.

person not necessarily by making his situation worse (although the dashing hopes or expectations may indeed do this), but by permitting the existing harm to continue when it could have been stopped. Such permitting constitutes acquiescence in the harm; it hence violates Davis's right to have his basic well-being respected.[15]

The two swimmers have equal rights to the necessary conditions of successful agency, and it is only Davis who sees the substance of this right in peril. He stands to lose his life. So when Carr helps Davis this cannot be an act of charity. In the case as described, Carr is an agent who is in full possession of his (basic) well-being, and being well-off as he is, he must assist Davis because he has a right to it. If he fails to provide that assistance, then by acquiescing in Davis's harm he performs an action that is contrary to Davis's equal rights.[16]

Unfortunately, there is a problem with the duty to assist — and with the duty not to send back — asylum seekers. The problem is that its fulfilment could obstruct our obligations toward far away refugees who did *not* make it to our shores. To see why that is, recall my general assumption that for any potential host country there is in principle only a limited number of entry places available. Note further that to act in accordance with the above established principle of *non-refoulement* will have the consequence that a certain part of those scarce places will be occupied by asylum seekers. From the perspective of a refugee who remains in a distant place, especially for someone who in a refugee camp waits for being *resettled* to a safer country, the fairness of the principle of *non-refoulement* would be hard to see. As Peter and Renata Singer put it, those other refugees will "wonder why if, by some miracle, they could set foot on the shores of a country like the United States, they would be entitled to jump the queue over their compatriots who had been patiently waiting for years for the chance of a place".[17] Michael Walzer also sees this dilemma:

> Why be concerned only with men and women actually on our territory who ask to remain, and not with men and women oppressed in their own countries who ask to come in? Why mark off the lucky or the aggressive, who have somehow managed to make their way across our borders, from all those others?[18]

In response to these questions, Walzer himself brings forward two general points to support the claim that it is nonetheless justified to help asylum seekers rather than all the distant others, or at least to

---

[15] *Ibid.*, pp. 224–25.
[16] See *ibid.*, p. 210.
[17] Singer and Singer (1988, p. 119).
[18] Walzer (1983, p. 51).

prioritize them. The first reason he evokes is that the alternative—namely not to help them—would involve the use of force against "helpless and desperate people".[19] This point seems to be a variant of what I have been proposing before, namely not to violate the basic rights of asylum seekers. His second reason builds on the assumption that the number of asylum seekers that reach a particular country is usually small. Under such circumstances of only few asylum seekers reaching one's shores, the costs of helping those people by taking them in would tend to be negligible, so that (and here he draws on the first reason) "we would be using force for 'things superfluous'".[20]

However, the way in which Walzer tries to solve the problems related with the *non-refoulement* principle remains unsatisfying. They are an expression of his intuitions rather than a well thought-out attempt to offer a substantial solution. Here is how the two Singers criticize Walzer's take:

> If there is any difference at all in the two cases, it lies in the symbolic significance of using force against an identifiable, harmless individual. We may rightly react strongly against the kind of callousness which can force people to go back to situations of great danger, but we should also react strongly against the kind of callousness which can leave people in such situations. If we take satisfaction in being too humane to deport someone seeking asylum while we continue to reject all applications from those in similar situations who have not made it to our shores, we are being hypocritical.[21]

What this objection suggests is that it is questionable to divide refugees into the few who have made it to our borders and the larger group of those who have not.[22] However, physical proximity is a more ambivalent criterion than the Singers admit. For after all, the mere presence of the refugee grounds a special relationship between refugee and host country: they are, as I pointed out before, in a transaction, just like the two swimmers Carr and Davis. And it is by the mere territorial presence of the refugee that one particular country is specified—at least in the concrete and immediate situation—as the natural candidate for granting assistance. On the other hand, and here the Singers make an important point, it remains somewhat arbitrary and presumably even irreconcilable with the principle of equal moral rights that we give priority to those who reach our shores (which may tend to be the

---

[19] Ibid.
[20] Ibid.
[21] Singer and Singer (1988, p. 120).
[22] Ibid.

stronger and financially more potent ones).[23] So under the condition of scarce entrance places the *price* for our special concern with the rights of asylum seekers could effectively be borne by other far-away refugees who are possibly ever needier. It thus appears to be questionable whether a country can consistently claim to take all refugees' rights to protection equally seriously while at the same time holding on to a practice in which asylum seekers may "use up" — to the detriment of other refugees — one of the most important resources there are to protect refugees, namely entry places.

Now, how should one deal with such (well-founded) reservations about the principle of *non-refoulement*? As I shall now contend, there are good reasons for holding on to that *principle*. States should hold on to it in full awareness that it may result in an unfair prioritization — on morally arbitrary grounds, that is — of some "lucky" asylum seekers over other more distant refugees. The decisive reason for my defence of the *non-refoulement* principle is a pragmatic one. It starts with the observation that most of today's potential host countries hardly portray a genuine effort and willingness of assisting (on a prioritized basis) those (allegedly) even needier refugees in distant countries. Arguably, it is only if countries credibly pursued this aim of first and foremost assisting those distant people who need it most, that the *non-refoulement principle* could be legitimately criticized on the basis that it compromises that effort of assisting the neediest. Otherwise, any critique of the *non-refoulement* principle that is directed at its arbitrariness and unfairness would lose its normative thrust. And this is presumably true for most countries in the world, so that I may here neglect that critique of the *non-refoulement* principle, justified though it may be in principle.

To underline this point, take the example of people fleeing the war in Syria: these people were left alone by the international community. Neighbouring countries could not (and some did not want to) shoulder the burden of taking care of large numbers of Syrian refugees and were shamefully left alone with that burden. At the end of 2014 it turned out that refugee camps in Jordan, Lebanon, and other countries were dramatically undersupplied. Donor countries largely ignored the appeal of the UN's World Food Programme (WFP) for further funds so that millions of refugees in the border region found themselves "facing a disastrous and hungry winter".[24] Many of the refugees affected by this shameful unwillingness to aid tried to make it to Europe. Now, the relevant question for the present discussion is this. Can we credibly

---

[23] See Hugo (2010, p. 26).
[24] Jones (2014).

send *these* people back by drawing on the rationale that to help *them* rather than other refugees (who equally need that help but remain in the border region) would be unfair? Of course, this is a rhetorical question. In contexts where little or nothing is undertaken to indeed help those distant refugees who are possibly even needier, it would be hypocritical to even question the *non-refoulement* principle on the grounds that it allegedly implies an unjustifiable privilege of the few refugees who were only audacious enough to leave the crisis region.

4.2.2. The Neglected Duty to Resettle

To the extent that in situ assistance was not possible or that it was for some other reason not realized as the preferable option in the concrete case, the international community of states and therewith all individual states face an obligation to make sure that refugees be assigned to a new state. They should be assigned to another state which, other than their former state, is capable of protecting them against threats that render life undignified. For any particular potential host state, the right of long-term and permanent refugees to be assigned a new state manifests itself as a duty to *resettle* refugees to its own territory. Resettlement is the second (and as I will contend *preferable*) of two channels by way of which states can and, as the case may be, should *admit* refugees (the first channel is the admission of asylum seekers, i.e. those who made it to the potential host country by themselves).[25]

There are good reasons for holding resettlement to be the preferable of the two immigration-related ways of assisting refugees. One reason is that it is doubtful for a country to accept a responsibility to assist the deprived while at the same time expecting *them* to incur the costs and to take the risk of long and dangerous journeys in order to get hold of that assistance.[26] This will in many cases mean that those whom a state ought to assist are not assisted after all, either because they do not venture to leave the crisis region or because they never reach their potential host state. In fact, for potential host states to wait for refugees to come to *them* (with some states hoping, we may presume, that they never actually turn up)[27] appears to be far from displaying a credible commitment of really assisting them.

---

[25] See Carens (2013, pp. 197–99).
[26] For the view that resettlement is the preferable (but neglected) way of assisting refugees, see also Kukathas (2016) and Owen (2016).
[27] For this presumption, see also Kukathas (2016, pp. 261–63). And for a more detailed disclosure and critique of the concrete strategies that most of today's richer states use to prevent refugees from arriving at their shores while at the

A further reason why resettlement should be the preferred option is that by resettling refugees countries express solidarity with those other potential host countries which for geographic reasons are disproportionately burdened with a given refugee population. However, the fact that of the more than 65 million[28] people displaced worldwide in 2015, there were only around 107,000[29] people that were submitted for resettlement means not only that there is a dramatic discrepancy between reality and that which is normatively demanded. It also tends to have two very tangible and deeply problematic real-world implications.

The first is that many refugees see themselves forced to embark on long and dangerous journeys to distant countries, provided they have the means to do so. Their journeys are long and dangerous, and in the absence of safe and legal routes, refugees will often have to pay for the services of the ruthless and illegal business of people smugglers. Many never reach the destination country.

The second morally inacceptable implication is that large numbers of refugees spend many years (or even decades) in refugee camps. They find themselves in what are technically referred to as "protracted refugee situations".[30] For the affected individual this means a most dehumanizing condition: she is caught in a condition where she possesses nothing but the most basic means for survival. That this situation is degrading becomes clear by drawing once more on the distinction between *occurrent* and *dispositional* freedom. People caught in such situations are not free in the long-term sense of the word. They cannot pursue their life plans let alone unfold their talents. How little they are free in the dispositional sense becomes clear already from the way the UNHCR defines a "protracted refugee situation": namely as a "long-lasting and intractable state of limbo".[31] Indeed, the relevance of the distinction between *occurrent* and dispositional freedom with respect to the case of protracted refugee situations is brought home fully by the UNCHR's observation that the lives of the affected people "may not be at risk, but their basic rights and essential economic, social and psychological needs remain unfulfilled after years in exile".[32]

---

same time formally upholding the right to asylum, see Dummett (2001, pp. 89–153).

[28] See UNHCR (2016, p. 2).
[29] See *ibid.*, p. 3.
[30] See UNHCR (2006, pp. 106–27).
[31] *Ibid.*, p. 106.
[32] *Ibid.* By a more technical definition used for UNHCR statistics, protracted refugee situations are those situations "where the number of refugees of a

Initially, their self-determined life was taken away by brutal states or made impossible by conditions under which their state could no longer provide for them. Trapped as they are in the limbo of protracted refugeehood, it will then appear to them that it is just as much the system of states that in their particular situation stands in the way of their ability to lead a self-determined life. For apparently they cannot just leave the camp and try to make a living in some other place of the world they happened to be born in. Such cases must be prevented. Regarding the 6.7 million refugees under the UNHCR's mandate caught in protracted refugee situations in 2015,[33] the international community of states has allowed for a shameful situation. The world's states must address this situation much more resolutely than they have, apparently, done so far. There can be no doubt that for the affected people the current arrangement of states and borders obstructs, rather than enhances, freedom.

After this case for the duty to resettle, the analysis will now turn toward the corrective justice-based claims to admission that some refugees could make in order to even strengthen their basic human rights-based entitlements to assistance.

4.2.3. The Duty (III-2) to Compensate: Causal Connection and Corrective Justice

A very powerful normative source for admitting refugees is given in those cases where a particular country played a relevant causal role in bringing about the reality of a particular flow or group of refugees. This normative source is complementary to the general human rights-based claims to admission which refugees can make regardless of why they became refugees. Now, there are at least two ways in which one can think of someone as being causally responsible for the harmful outcome of refugeehood.

(1) For one thing, there may be cases in which a party (i.e. in our case a country or its state) is held responsible for certain specific actions that contributed to harmful effects suffered by others, for example their displacement. The kind of causal connection I have in mind here is one that allows us to put the finger on one or more clearly identifiable agents whom one can hold responsible for their performing an action that is *contrary to another person's or group of persons' rights*. This possibility of tracing one or more particular responsible parties then allows

---

certain origin within a particular country of asylum has been 25,000 or more for at least five consecutive years" (*ibid.*, p. 107).

[33] See UNHCR (2016, p. 20).

us to rely on the normative principle of corrective justice as introduced earlier. In such instances, one can put the fingers on the perpetrators, and knowing that with a view to a widely accepted baseline (i.e. the "normal") they acted in a disrespectful way, one may rightly hold them liable for remedying what they helped bring about, namely a particular refugee population. By isolating some particular agents, one can distinguish them from others who are by implication not responsible.[34] For example, Norway was not responsible for Vietnamese refugees in the same way that the United States was.

This example of Vietnamese refugees leads us to a much quoted passage in Michael Walzer's treatment of refugees. Walzer asserts that we have a special responsibility toward people whose refugeehood we brought about. "The injury we have done them makes for an affinity between us: thus Vietnamese refugees had, in a moral sense, been effectively Americanized even before they arrived on these shores."[35] The responsible parties have a duty to make whole whatever negative circumstance they helped bring about. We have a duty to compensate for our wrongdoing toward others by granting them a new home in our country in so far as this is the most effective or the only remaining option by way of which we can restore what was through our own actions taken away from them. This verdict flows naturally from the account of corrective justice as proposed earlier in this work.

Drawing on the same example of special American duties toward people from Vietnam, Laos, and Cambodia (or, by the same rationale, toward Iraqi and Afghan refugees today), Joseph Carens similarly assumes that we are obliged to admit refugees when "the actions of our own state have contributed in some way to the fact that the refugees are no longer safe in their home country."[36] Where such a causal link between the actions of our state and the plight of others can be established, it functions as a normative source that adds to the more general reasons to assist refugees as discussed above. In cases where without a particular country's actions the harm leading to displacement would not have come about, i.e. cases where one can identify some particular countries' actions as a *sine qua non condition* of the displacement-triggering event, it would amount to an absolute injustice if the thusly responsible country, after failing to assist in situ, denied to admit the affected refugees.

---

[34] See Young (2011, p. 105).
[35] Walzer (1983, p. 49).
[36] Carens (2013, p. 195).

(2) For another thing, there may be instances where responsibility is not that easy to attribute and where blameworthy parties cannot be readily identified and isolated. This seems to be the case with a view to *structural injustices* to which many (or potentially all) parties contribute — presumably to varying degrees. One concrete example for such a structural injustice is globalized economic *processes*, to which innumerable actors contribute in untraceable ways. Such processes may cause harm to some and force them to leave their homes. Another example is climate change, which causes harm and to which all people around the globe contribute through emissions of greenhouse gases. They all contribute to the problem, so in some way they all contribute to the harm caused by global warming, and they all — though undoubtedly to largely varying degrees — contribute to the migration and forced displacement that is being caused by such human-induced climatic processes. The problem with this second type of causal connectedness is (i) that the injustices are the *cumulative* result of the actions of many or all parties so that no countries' causal contribution amounts to a *sine qua non condition* of the harmful outcome in question, (ii) that responsibility cannot be isolated, and (iii) that the individual actions — whose cumulative effects may cause harm and constitute blatant injustices — are by themselves in accordance with widely accepted social rules, conventions, practices, and policies (so that they do not deviate relevantly from the "normal"). With varying connotations, this general kind of injustice may be referred to, for example, as "New Harms"[37] or simply as "structural injustice".[38]

The differences between the two kinds of causal connectedness are clear. While there is in the current academic debate on the morality of immigration a near consensus that the first kind of causal contribution can render an already strong duty to assist even stronger, there is less clarity on how to deal with the second kind. In view of the global pervasiveness of structural injustices, this gap in the current literature constitutes a serious shortcoming.

Note that at this stage I will not yet propose ways for dealing with this second, more blurry type of causal connection. However, a more systematic way of dealing with blurry causality is provided in my later discussion of the peculiarities of climate migration. In that later discussion, I will not only explain why such causal uncertainty is especially accentuated and problematic in the case of climate migrants. I will also develop an argumentative strategy by way of which that

---

[37] Lichtenberg (2014, pp. 73–90).
[38] Young (2011, p. 106).

issue of causal blurriness could be sidestepped so that the affected climate migrants would in principle retain the possibility of making claims to assistance and/or admission under corrective justice. Possibly, other non-climate refugees who are equally the putative victims of causally blurred structural injustices could amend that proposed argumentation for their purposes.

4.3. Distributing Refugees

A short look at the current distribution of the world's refugee population makes one thing very clear: there is no functioning and no fair distribution of worldwide refugees. In 2015, 86% of the world's refugee population was hosted by Developing Countries, and even the Least Developed Countries granted asylum to over 4 million refugees.[39] These countries are in a relatively weak position to provide the means for an adequate livelihood and they stand a relatively great risk of being overburdened. While it is easy to see that such distribution among the world's states is grossly unfair, the more difficult question to be addressed in the following sub-chapter is what a fair allocation of refugees would look like: what are the criteria on the basis of which to determine which country should take in what share of global refugees? This is an important question because the responsibility that arises from the existence of a global refugee population in need of assistance potentially falls on many if not all states and has to be divided among them by some criteria. The responsibility is not *indivisible* in that situation,[40] as it is, for example, in the case of Carr, who is effectively the only party that qualifies as a potential helper.

Assuming full-compliance of the world's potential host states (at least until I later take into account the reality of partial or non-compliance), I shall now propose five criteria for the distribution of refugees. Note that these five criteria are closely interrelated, that many of them resort in one way or another to aspects that were already discussed in the previous chapter on the morality of immigration (like the importance of functioning economic, political, and cultural institutions), and that the order in which I shall present them here does not indicate their relative importance. Rather, the concrete weighing of these criteria, which would clearly be necessary in order to formulate the concrete terms of a fair refugee allocation, is highly context-dependent and is therefore not attempted here.

---

[39] See UNHCR (2016, p. 18).
[40] See Miller (2016, p. 35).

(1) A first criterion that should play an important role is *proximity* of a given state to the crisis region. This may be surprising at first glance, for geographic proximity is indeed a contingent fact. And yet, there are good reasons why (alongside the other criteria) it should inform the way in which refugees be distributed among states. At least as an immediate response to a refugee crisis it would be necessary that the geographically close countries take in a larger share of refugees, even if the reasons for this extra burden are largely logistical in nature. The aim is that refugees get relocated as quickly as possible, but to achieve that aim it is organizationally reasonable that they first find shelter in neighbouring countries—to which they are of course not so much "allocated" as that they flow there rather naturally. At any rate, the principle of *non-refoulement* seems to make unavoidable that proximity plays that central role at least with a view to the first phase of allocation. Fleeing to these neighbouring countries, refugees must not be sent back and must find (a first) shelter there.[41] Of course, these countries should be assisted by the international community. It is in distant countries' interest that refugees find assistance in neighbouring countries, not only because assisting them there tends to be cheaper but also because the number of refugees that will try to make it to those more distant countries is *ceteris paribus* reduced in this way. As Betts and Collier point out, refugees that remain in neighbouring (developing) countries should be enabled to work and participate socio-economically there.[42] Provided the international community is willing to assist effectively, empowering refugees in the region can be a chance both for the affected refugees and for the assisting developing country. A final reason in favour of assistance in the border region is that refugees that remain geographically close to their home country are more likely to return and rebuild their country once the conflict is over —an aspect that is widely neglected in normative discussions on refugee policy.[43]

(2) As I contended previously, the fact that a particular country or state bears a *causal responsibility* for the existence of a particular group of refugees puts this country under extra normative pressure to assist that refugee group. This consideration should not miss in the present account either. It figures in this account as an important criterion for the fair distribution of refugees. Indeed, it would be grossly unfair to have one country bear (disproportionately) the costs of dealing with a

---

[41] See Carens (2013, p. 213).
[42] See Betts and Collier (2017, especially chapters 6 and 7).
[43] *Ibid.*

refugee population even though *another* country caused it or, as concerns structural injustices, even though that other country contributed relatively *more* to the displacement-triggering outcome. In such cases where countries share causal responsibility for a given group of refugees, relative contribution would be an appropriate criterion that should inform any attempt to specify what a fair allocation of a particular group of refugees (or of all refugees) means in the concrete case.[44] In any case, causal contribution is a relatively weighty criterion: to the extent that some countries contributed to refugees' deprivation of something they are entitled to—namely their secure livelihood—such causally contributing countries have a duty to make whole the affected people, which in practice will materialize as at least a more burdensome share of the refugees who are being distributed. I assume that in cases where individual countries' actions can be singled out as a *conditio sine qua non* of a particular displacement-triggering event, the criterion of causation (which in such cases is not merely causal "contribution") would be the all decisive one. The duty of the United States to take in people who were forced to leave the Marshall Islands after the US hydrogen bomb test in 1954 is certainly as strong as a duty to admit can get. Such a decisive causal role would dwarf the relative weight of all other criteria adduced to allocate a particular group of refugees.

(3) The third very important criterion combines a whole range of factors that I shall here unify under the term of "*absorptive capacity*". By this I shall refer to such factors as financial resources, economic capacity (for example that a vibrant economy can integrate into its workforce more refugees than a stagnant economy), territorial and environmental capacity, and social capacity (in the objective sense of whether or not institutions such as the education system, social welfare, and the health care system are stable enough and have sufficient capacities to support further people). All of these aspects are of course interrelated and overlapping. As I contended at an earlier stage, it is morally legitimate to take these factors seriously when deciding on a morally acceptable immigration policy and it stands to reason that they may then also influence what is assumed to be a fair allocation of refugees, i.e. one that burdens no country in an unreasonable and disproportionate way. Note that this concern with the *absorptive capacity* of host states is ultimately grounded in the same rationale that justifies the assistance granted to refugees: that people's basic rights ought to be protected. If states are overburdened with large numbers of refugees to the extent that they can no longer adequately protect the basic rights of

---

[44] See Owen (2016, p. 282) and Carens (2013, p. 213).

their own citizens, then such assistance to refugees would turn out to be self-defeating. The issue of absorptive capacity is also related to the next criterion of cultural proximity.

(4) *Cultural proximity* between the host society and the refugee group is a further relevant criterion to be taken into account. In the previous chapter I unfolded the view that culture matters and that it can be assigned a systematic place in rights-based theory in so far as it provides individuals with the valuable good of a secure *context of choice*. More specifically, I argued that culture is important when discussing immigration policy. Relying on robust social scientific evidence, my contention was that the rapid and large-scale influx of immigrants, especially where they are culturally distant, tends to threaten or at least reduce existing levels of social trust among citizens. In my previous argumentation, this recognition mainly served to support the view that the exclusion of non-refugee outsiders is morally justifiable. It now recurs as a consideration that should inform our answer to the question of how to *allocate* refugees (but not on whether or not to assist). If the mentioned negative effects of admitting an overly large number of culturally distant people can be avoided by allocating them preferably to other countries where less cultural friction is to be expected, then (i.e. provided such "other countries" exist) there seems to be a strong case for doing so.

As a matter of fact, there can be no doubt that fears of cultural alienation figure centrally in public debates and, unjustified though they may be in some cases, they should be taken seriously. To weigh this criterion of cultural proximity carefully against the other ones could help maintain among the members of the host society higher levels of commitment to the existing refugee regime, which after all is an important moral goal governments should pursue. However, to what extent this criterion ultimately should play a role depends on the number of refugees to be distributed, for only where numbers are large would my quarrel with cultural proximity become an issue in the first place. It also depends on the overall social and economic context of the potential host country: it is to be presumed that where the economic situation is dire and where jobs are scarce for the existing population, the risk of xenophobic backlashes would be higher.

(5) When reasoning on a fair allocation of refugees between states, what should also be taken into account is the *preference of the refugees* themselves. As Joseph Carens observes: "They are not just victims to be assisted in whatever way the receiving states deem best. They are

human beings whose agency deserves respect."[45] The allocation of refugees should thusly be fair not only from the perspective of host countries, as most of the previous criteria seem to suggest, but also from the perspective of refugees. They may, for example, have family members in a particular country, which would provide for a very strong reason for them to be allocated to that country. Only in situations when too many refugees express a preference to be sent to a few especially popular countries would it become difficult to fulfil that preference without making some sacrifice on any of the other mentioned criteria. So again, what is required is a fair and context-dependent weighing of the different criteria.

Having established criteria for a distribution of the world's refugees, it now bears taking into full account that in our imperfect world far from all states can be assumed to *comply* with whatever distribution is proposed. So how to deal with the reality of partial compliance? The response to this question is straightforward: those states that *do* comply need to take up the slack for other non-compliant states.[46] To see why this stance resonates with the present work's human rights standard, an analogy may be adduced: assume two men walk by a pond and see two children drowning who have been pushed into the pond by a local villain. The two men quickly distribute the burden of helping and decide that man A rescue the little boy and man B rescue the little girl. They distribute the burden in this way not only for the sake of helping as quickly as possible and because this is what *human rights* require, but also because, being equally in a position to help, *fairness* requires this division of the burden to assist. After man A did his fair share of rescuing the little boy, he realizes that man B refused to do his share of rescuing the little girl. Although the little girl has not yet fully drowned and although man A is capable of rescuing this second child, he decides *not* to rescue the little girl. This analogy is meant so show that it would be clearly wrong not to take up the slack when others fail to do the morally right thing. Just as A would be obliged to also assist the girl, compliant states are obliged to also assist "surplus" refugees. In other words, they are obliged to assist regardless of what other states do. The concern with human rights is lexically prior to other principally relevant normative considerations like fairness: it may amount to an *unfairness* that man A must go beyond his own presumed fair share of helping only one child, but the human rights at stake on the side of the second drowning child clearly outweigh this injustice suffered on man

---

[45] Carens (2013, p. 213).
[46] For a recent argument in support of this view, see Owen (2016).

A's side. Failing to help the girl is rendered not even marginally less serious a moral wrong through the fact that someone else was originally supposed to help that child—this follows from the principle of taking human rights' to be lexically prior.[47] All that matters in the situation is that the girl sees the substances of her basic rights acutely threatened and that man A is *still* able to assist at no comparable cost. Likewise, a country is obliged to assist refugees until it is no longer capable of doing so at no comparable cost, largely regardless of other states' conduct. The decisive question is then this: when is it that countries can reasonably claim to be no longer able to assist at no comparable cost? When does their duty to assist refugees reach its limit?

In a world of non-compliant states as the present one—a world in which some states might indeed be confronted with very large numbers of refugees—it will be all the more important to know where these limits are and how they can be specified and justified. This specification and justification of the limits of our obligations toward refugees is the task of the next sub-chapter.

4.4. The Limits of Our Duties to Admit Refugees

Depending on the number of refugees who make a request for asylum, compliant states could at some point lack the resources for taking in further refugees. Clearly, refugees make a very powerful claim to admission, but all the previously discussed reasons that speak in favour of a right to exclude in principle hold in the context of refugee policy just as well. To be sure, when confronted with the strong claims of refugees, potential host states will be urged not to *exert* their right to exclude if that exertion can be avoided—they will be urged by the recognition that they have a duty to assist those needy people. But there are limits.

To have them take in ever further refugees only to see those heroic states eventually collapse under the burden of too many refugees would be absurd. At this endpoint economic, social, or political institutions—which as I argued depend on complex schemes of social cooperation and which are often only locally stable—stand the risk of breaking down, potentially with the self-defeating result that *more*

---

[47] See also Karnein (2015), who argues convincingly for the duty to take up the slack. Similar to the point I have been making here, she contends that the question of what compliant parties owe to needy third parties ought to be kept separate from the question of how to deal with the unfairness that the non-compliant parties do to the compliant parties. So the injustice done to the compliant swimmer should have no bearing whatsoever on the question of what the compliant swimmers owes to the drowning persons.

people would find their basic rights unprotected than would be the case if the state had at some point begun to *refuse to protect* the basic rights of further refugees by taking them in. In terms of overall human rights protection, nothing would be gained then. This suggests that it is in principle conceivable for a state to exclude refugees at some point without giving up its claim to take human rights seriously. As already argued, states have a right to exclude refugees for the same reason that they have a right to exclude migrants: they need that right in order to protect the institutions necessary for securing the citizens' rights. They must have it so as not to lose control over the functioning of vital institutions. All else would be incompatible with the citizens' right to self-determination. It is in this sense that the citizens of a state may decide on where to draw the limit to the number of refugees admitted. The present sub-chapter is concerned with the question of how states can exert that decision in a morally acceptable way. When does the duty to admit refugees reach its limits?

As a first approximation and as already suggested, one can say with certainty that there can be no duty to help beyond that point where one can no longer help at no comparable cost, because to help beyond that point would conflict with the here upheld notion of equal rights. To assume a duty to assist others in securing their basic rights even when such assistance would mean to risk one's own basic rights is irreconcilable with the assumption of equal rights. It would imply that the assisted refugees have more of a basic right to some good than the helper herself, which would be inconsistent with the assumption of the equal rights of persons.[48] The ultimate point beyond which assistance to refugees by granting them shelter can no longer be demanded meaningfully is reached when the act of offering that assistance assumes a self-defeating character. Holding on to the notion that a state has the primary duty (II) to protect its citizens' basic rights, a state cannot have an obligation to assist refugees if doing so seriously compromises its ability to protect that primary duty (II).

The citizens of the potential host state, however, can hardly specify when this ability begins to be compromised. They do not know how, when, and whether this point is ultimately reached; it remains an uncertainty. So how can one deal with this uncertainty? What does it imply for the issue of setting limits to our obligations toward refugees?

In a recent philosophical essay on the question of where to draw such limits, German philosopher Matthias Hoesch (2016) recognizes that the actual consequences of the further immigration of refugees are

---

[48] See Steigleder (2016, p. 254).

difficult to assess. Based on this recognition that the real consequences for the social order within the host state are *uncertain*, he contends that a concern with (such "merely" potential) negative consequences on social order should have little normative bearing on the question of whether and when to set limits to the number of further refugees admitted.[49] The rationale Hoesch displays here is indeed a most disconcerting one: in his view, the *certainty* that refugees will be (or will remain) deprived of the substances of basic rights when denied admittance dwarfs what is "merely" the *risk* of overly strained social institutions (which could then leave some citizens without adequate protection of their substances of basic rights). The implications of this rationale are arguably inconsistent with the assumption of the equal rights of the existing citizens. Hoesch reveals here a problematic insensitivity to the relevance of risks in the immigration context.

That risk should be a central concern when debating immigration policy was already suggested in earlier discussions. Recall that social cooperation games that underlie social institutions are only locally stable. It is against the backdrop of this recognition that Collier warns that "beyond some unknowable point the losses from reduced mutual regard [from admitting immigrants on a large scale] are liable to increase sharply as thresholds are crossed at which cooperation becomes unstable. Cooperation games are fragile because if pushed too far they collapse."[50] To the extent that this claim is empirically valid, it holds in principle as much for the case of refugees as it does for other non-refugee immigrants. What may be uncertain then is the question of *when* the point is reached beyond which cooperation games become unstable. But what is clear enough is *that* such a point exists. It is precisely this combination, i.e. of *knowing that* there is that point that must be prevented and of *not knowing where* exactly it is, that urges us to rely on a risk-ethical perspective.

It is crucial to recognize that no host state should wait until that point is reached when cooperation is virtually on the verge of collapse

---

[49] See Hoesch (2016, p. 28). Hoesch then goes on to contend that a country must indeed risk a breakdown of its social order provided the neighbouring countries (through which many refugees now pass) have already taken in so many refugees that their own social order has equally reached the point of breakdown.

[50] Collier (2013, p. 63). For a valuable discussion of the importance of cooperation and the great challenge of maintaining it, see Haidt (2012) and Greene (2013). As Greene puts it trenchantly, "[f]rom simple cells to supersocial animals like us, the story of life on Earth is the story of increasingly complex cooperation. Cooperation is why we're here, and yet, at the same time, maintaining cooperation is our greatest challenge" (Greene, 2013, p. 59).

and act only then. Rather, the relevant point is that at which cooperation is rendered unstable, and that point is reached (presumably much) earlier than the point at which cooperation is on the verge of actual collapse. This decisive claim is informed by the assumption that, as Collier points out, it may prove difficult to restore the stability of cooperation games once they become unstable (and it is only within this *phase of instability* that there is a serious risk of collapse in the first place).[51] The relevant risk one should be concerned with is thus not primarily the risk that cooperation and public order *actually* collapse, but the risk of reaching that unknown tipping point at which cooperation becomes unstable.

Now, on the basis of this refreshed social scientific understanding, I will come back to my above critique of Hoesch: namely that in his discussion of where and how to define the limits of our obligations toward refugees he neglects the relevance of the risk-ethical dimension. To substantiate this critique, consider how Hoesch draws on the pond-analogy, with one swimmer (or rescuer) and many persons drowning. In his discussion of the swimmer's obligation to rescue further persons Hoesch introduces a helpful distinction between two potential successive *phases* of the rescuer's health situation: first he could catch a (serious) cold from swimming in the cold water, and second this cold could turn into a life-threatening pneumonia.[52] Arguably, either of these two phases may be seen to parallel the two just discussed phases: the serious cold parallels the *phase of instability* (which I identified as the relevant risk), and the life-threatening pneumonia parallels the phase at which cooperation is on the verge of actual collapse. Now, the crucial problem with Hoesch's account is that he does not conceive of the serious cold as the relevant risk though. Instead, he is concerned mainly with the pneumonia-risk. On Hoesch's account, the swimmer may well catch a cold. It is only then that measures should be taken to prevent the cold from turning into a life-threatening pneumonia. This assessment is irresponsible.

Contrary to Hoesch, I contend that the risk of a life-threatening pneumonia is a risk that must be avoided in the first place. (It is, in other words, not primarily the pneumonia that must be prevented, but already the *risk* of suffering such a life-threatening pneumonia and this implies that the cold must be prevented.) In so far as the stakes are high and precisely because we have only limited information on the actual consequences of taking in ever further refugees, we are urged to *tread*

---

[51] See Collier (2013, p. 63).
[52] See Hoesch (2016, p. 28).

carefully and to choose an approach to the question of limits that is maximally sensitive to risks.

Of course, it must be made sure that such an approach that is sensitive to risks takes into equal consideration the risks imposed on all affected sides. But whatever impartial risk-ethical perspective is warranted in the field of refugee policy, it is self-limited by the reasonable assumptions that states have the primary duty to protect their own citizens, and that those very citizens cannot be obliged to do something that could turn out to be self-defeating. Thus, a state may set a limit to the number of refugees admitted even when, or precisely because, it is uncertain how probable or improbable the materialization of the risks associated with admitting them is. This uncertainty marks the reason why a risk-ethical perspective is needed to define limits in the first place. The state, just like the swimmer, hardly knows when exactly it incurs a serious risk of a life-threatening pneumonia, and this uncertainty suggests what should be the primary objective: not to catch *a cold* in the first place. But there is a problem with the analogy of the swimmer and the state: while the swimmer could of course—in the sense of a supererogatory act—decide to impose upon himself that risk of pneumonia by rescuing further people, it is most doubtful whether the citizens of a state may democratically decide to take that risk of pneumonia. Should the risk materialize, it would also harm those members of society that could not participate in (or voted against) the decision to take the risk. For the state to take the risk of pneumonia would therefore appear to be a most doubtful thing to do.

Based on general indications or first symptoms already triggered through the cold water, the swimmer may at some point decide to stop rescuing further people. He cannot but make that decision under uncertainty, so what he has to work with are rough indications rather than unambiguous data and hard facts. The same holds true for democratically elected governments. When they have indications that the risks to the stability of their institutions become too high, they should avoid the incurrence of further risks. It turns out then that the oft-adhered to verdicts of "ought implies can" or "ultra posse nemo obligatur" tend to be oversimplistic in the context of immigration and refugee policy. Such phrases suggest clarity where in fact there is uncertainty: the potential helpers will have problems specifying to what extent they actually *can* help. Whether they can and thus ought to help is a matter they ultimately cannot but couch as a risk-ethical question.

Part II

Debating Immigration Policy in the Climate Context

Five

Facing a Heated Reality
Climate Change and Migration

Human influence on the climate system is clear, and recent anthropogenic emissions of greenhouse gases are the highest in history. Recent climate changes have had widespread impacts on human and natural systems.[1]

Changes in many extreme weather and climate events have been observed since about 1950. Some of these changes have been linked to human influences, including a decrease in cold temperature extremes, an increase in warm temperature extremes, an increase in extreme high sea levels and an increase in the number of heavy precipitation events in a number of regions.[2]

Most individuals will suffer from climate change, although not uniformly and not in any proportion to their contribution to causing climate change. It is perhaps unfair simply that the benefits have been narrowly held while the costs have been widely disbursed. It is certainly deeply unfair that the benefits have been narrowly held by those who have inflicted the damage on everyone while the costs, including severe harms—loss of life, health, or home—are falling randomly upon all.[3]

The ongoing accumulation of greenhouse gases in the earth's atmosphere over the past 250 years has severe impacts on the climate system. It is getting warmer, which goes along with deep effects on the biosphere, on the shape of the earth's surface (e.g. through sea-level rise, melting glaciers, etc.), and therewith on the very environment on which we all depend for our lives. Climate change has an impact on the conditions under which we humans try to live together. The natural environment is the context on which our forms of social cooperation depend, so that a changing environment will invariably affect social organization. I take the science behind climate change to be clear, which allows me to make two hardly contestable assumptions: the

---

[1] IPCC (2014b, p. 2).
[2] *Ibid.*, p. 7.
[3] Shue (2015, p. 11).

occurring climate change is predominantly man-made, and it is a threat to human civilization. Through the versatile ways in which vital human interests are (actually and potentially) affected, anthropogenic climate change is then clearly (also) a human rights issue. There can be no doubt that rising sea-levels or harvest losses from increased aridity will, all else being equal, tend to decrease human freedom and compromise the autonomy of many people around the world.

One way in which this loss in freedom will play out is that people (are forced to) leave their homes in response to changes in their natural and social environment. At the same time it is this migration that allows people to adapt to the climatic forces they have become victims of. At least for some, migration is the chance they can grasp to recover (or retain) their autonomy and secure the basic goods they have a human right to.

The importance of making climate-induced migration an object of moral discussion is thus beyond doubt. Clearly, it would be unsatisfying for such a discussion to focus only on the victims' side—as if climate change were just a neutral process that hits some people who are unfortunate enough to live in the wrong regions of the world. A complete picture must take into account that there tends to be some gross unfairness involved in the reality of anthropogenic climate change where, as Shue puts it in one of the quotes opening this chapter, "the benefits have been narrowly held by those who have inflicted the damage on everyone while the costs, including severe harms—loss of life, health, or home—are falling randomly upon all."[4] This injustice must enter the moral debate on climate migration as an important background assumption. In any case, this assumption shows why the debate on immigration heats up once climate enters the discussion as a further variable.

It is increasingly the case that people's carbon emissions in one place of the world can have (negative) effects on distant people in other places. The way in which people are connected remains indirect and nebulous though, and whenever there are effects on others they will be the result not of attributable individual emissions but of an accumulated total so that the causal influence of individual emissions becomes untraceable. As Stephen Gardiner points out, "the impact of any particular emission of greenhouse gases is not realized solely at its source, either individual or geographical; rather, impacts are dispersed to other actors and regions on the earth."[5] In this way, the process of

---

[4] Shue (2015, p. 11).
[5] Gardiner (2010, p. 88).

climate-induced migration brings together people who were once distant strangers. For some high-emitting countries that so far remained largely unaffected by the negative consequences of climate change, it may be the masses of climate-induced migrants that could finally bring home even to them the insight that recklessness has costs, not only for others.

How useful are my previous elaborations on the ethics of immigration for the issue of climate-induced migration? Do the normative results on the legitimacy of immigration restrictions hold for the case of climate migrants? The intuitive assumption is that in some sense "we" are causally involved in the reality of climate migration. This appears to make climate-induced migration relevantly different from the other discussed forms of migration and it appears to call for a different (and supposedly stronger) form of moral responsibility toward the affected migrants. What is morally special about climate migrants? And what is it that we morally owe to them?

5.1. The Climatic Drivers of Migration

> Climate change is projected to increase displacement of people (medium evidence, high agreement). Populations that lack the resources for planned migration experience higher exposure to extreme weather events, particularly in developing countries with low income. Climate change can indirectly increase risks of violent conflicts by amplifying well-documented drivers of these conflicts such as poverty and economic shocks (medium confidence).[6]

The ways in which climatic changes could induce human movement are versatile. In order to give an account of this heterogeneity, I will draw on a typology of climate-induced migration as proposed by legal scholar Walter Kälin.[7] He identifies the following five *scenarios* by way of which people could be turned into climate migrants:

(i) *Sudden-onset disasters* (e.g. *events* like hurricanes, flooding, mudslides, etc.), which tend to generate temporary rather than permanent migration. Only relatively few people displaced by sudden-onset disasters will cross international borders.[8]

(ii) *Slow-onset environmental degradation* (e.g. processes like salinization of soil through rising sea-levels, droughts, water scarcity, etc.), which is mostly associated with *permanent* migration. To what extent migrants affected by slow-onset processes will move

---

[6] IPCC (2014b, p. 16).
[7] See Kälin (2010, pp. 84–92).
[8] See also Barnett and Webber (2010, p. 40) and Brown (2008, p. 17).

internationally will depend among other things on the availability of (economic) opportunities *within* the affected country.

(iii) *Sea-level rise* (SLR) and, in consequence, drowning island states constitute a special case of *slow-onset*-environmental degradation. SLR leads to clear cases of permanent migration. Most people will leave long *before* the islands are literally submerged by the rising tides, which leaves us with the key conceptual question of whether people leaving *in anticipation of* such ultimate submersion should be seen as *voluntary* or *forced* migrants (in other words: here the distinction between voluntary and forced movement would appear to depend entirely on the "timing factor" of people's movement).

(iv) Governments evacuating people from territories designated high-risk zones in terms of environmental dangers.

(v) Public unrest or conflict in response to decreasing resource availability.

Against the backdrop of this typology, it is worth taking a more detailed look at the concrete climate-induced events and processes that trigger migration. While the structure of my following overview follows Kälin's typology, it slightly diverges from it in so far as the scenario of planned evacuation (iv) plays no role in it and in so far as it subsumes the special case of SLR under the more general rubric of *slow-onset* triggers. Needless to say, affected populations may become victims of a combination of the following migration-triggering factors.

5.1.1. Sudden-Onset Triggers

Changing intensity and changing frequency of freak weather: The IPCC projects that, because of climate change, extreme weather events like tropical cyclones, storms, heavy rains, and floods are bound to increase in number and intensity. There is empirical evidence for such increased frequency in the past, and the higher intensity can be credited to warmer surface temperatures of the oceans as a corollary of warmer air temperatures. To put it more generally, there is more energy in the climate system, and more intense storms are one manifestation of such increased energy.[9] Moreover, it is in conjunction with sea-level rise that cyclones will become more intense and more destructive as storm surges will drive larger masses of water inland, potentially leaving

---

[9] See Rebetez (2011, p. 44). See also Emanuel (2005, pp. 686–88), who finds in his study that hurricanes, getting increasingly destructive, have more than doubled in number since the 1970s—a process he traces back to higher tropical sea surface temperature.

affected lands inundated and devastated.[10] The number of people yearly affected by storms and floods is already high (138 million between 2000 and 2008),[11] but it remains difficult to predict *to what extent* a higher frequency of such disasters will be reflected in the magnitude of human populations affected. While it can be said with certainty that such *sudden-onset* phenomena will cause displacement, there remains uncertainty as to the kind of movement triggered. There is disagreement on whether the triggered movement will be predominantly short-term and within the borders of the affected countries or whether it will also be permanent and across borders. In any case, the negative effects of more frequent freak weather on human populations will be highly dependent on the region where such events occur. The more a certain community or its individuals depend on their natural environment to make a living, the more they will be affected by disastrous events, whose impact would then be further exacerbated where it meets with local poverty. In such vulnerable localities, regions, or countries there are likely only inadequate means to (proactively) protect against such risks.[12]

This last recognition leads me to two related concepts that will be essential for conceptualizing and categorizing climate migration in the following sections: *vulnerability* and *adaptive capacity*. Although I now introduce them in this section on *sudden-onset* disasters, they are just as relevant for all other forms or triggers of climate migration: in the context of climate migration, *vulnerability* comes into play as a decisive social dimension. It determines the extent to which an individual, a household, or a community is forced to move in the face of a given detrimental environmental process or event. *Vulnerability* shall be understood here as:

> [T]he degree to which a system is susceptible to, or unable to cope with, adverse effects of climate change, including variability and extremes. Vulnerability is a function of the character, magnitude and rate of climate change and also the extent to which a system is exposed, its sensitivity and its adaptive capacity.[13]

Although it is itself part of the function of vulnerability, it is worth highlighting separately the *adaptive capacity* of a (human) system. It refers to a system's ability "to adjust to climate change (including climate variability and extremes), to moderate potential damage, to

---

[10] Doyle and Chaturvedi (2011, pp. 280–81) and Rebetez (2011, p. 44).
[11] See Piguet *et al.* (2011, p. 6). They draw on the data of the International Disaster Database EMDAT (www.emdat.be).
[12] See Kniveton *et al.* (2008).
[13] Houghton (2015, p. 163).

take advantage of opportunities or to cope with the consequences".[14] When vulnerable communities lack the capacity to adapt in situ to certain environmental events or processes, then a remaining form of adapting to such changes could be to leave the affected place. I refer to this adaptive strategy as "migration as adaptation".

5.1.2. Slow-Onset Triggers

Changing temperatures: The rise in global average temperatures is presumably the consequence that is most readily associated with climate change; it is the main parameter reflecting higher concentrations of greenhouse gases in the atmosphere, and many of the other (migration-triggering) consequences to be listed below are themselves a consequence of rising temperatures. But already by itself, higher temperatures must be seen as a stressor that can worsen human living conditions in some areas in such a way as to induce the affected people to migrate.[15] Rather than by the rising average temperature itself, human communities will be affected most severely by the increased frequency of hot extremes: heat waves have the potential to kill many, as was the case even in Europe in 2003.[16] As a more indirect consequence, heat waves tend to come along with drinking-water problems due to higher demand, reduced water quality associated with warmer temperatures, and increased rates of evapotranspiration, all of which will tend to stress human populations and their conditions for agriculture.[17] Where such extremes occur with increased frequency, people could at some point come to regard their situation as too precarious and therefore consider leaving. Concerning the rise in *average* temperatures, there are further indirect migratory triggers one should pay attention to: while rising temperatures may increase crop yields in colder regions, they will tend to cause decreased yields in (already) warm regions.[18] Moreover, an increasing number of insects associated with higher temperatures will put some regions under additional migratory pressure, either via the effects on human health (e.g. malaria) and public health systems.[19]

---

[14] Ibid.
[15] See Leighton (2011, pp. 331–40).
[16] See Houghton (2015, p. 200).
[17] See *ibid.*, pp. 184–87. Another disquieting account of future water challenges and of how they relate to migratory issue is provided by Cronin *et al.* (2008).
[18] See Rebetez (2011, pp. 38–41).
[19] For the nexus between climate change, health, and migration, see Barbieri *et al.* (2011), Carballo *et al.* (2008), Barnett and Webber (2010), or McMichael *et al.* (2012).

Melting glaciers: The rising of temperatures will cause the continuous shrinking of glaciers. Especially in the Andes, the Indian sub-continent and in some Chinese regions, such melting could unfold its negative consequences for human populations in at least two different ways.[20] First, it may cause flooding especially during the decades in which it is about to melt down. As a more extreme flooding scenario, there is even the possibility of a glacial lake outburst flood, which might require the evacuation of human communities. And the second way in which melting glaciers could unfold negative impacts might occur during dry seasons. With the glacier getting continuously smaller, water supply for people depending on the glacier's drinking water may decrease.[21] Of course the effects of such reduced water supply on human communities in the valleys will depend on other parameters like precipitation changes through climate change in the affected region. As for the Peruvian Andes, rain will with some likelihood decline as well. Such reduced water supply could put communities under pressure, especially if they grow in size.[22] Migration may become the natural safety valve then.

Changing precipitation and drought incidence: It is expected that global precipitation patterns will change significantly as a consequence of climate change.[23] While some areas will get more and heavier rain, other regions are likely to experience a drier climate, which will be felt hardest in those places where rain is already rare: as concerns sub-Saharan Africa, projections estimate a reduction in annual rainfall of 10% by 2050, and in other regions droughts are likely to become more frequent.[24] On a global scale, changing precipitation patterns in conjunction with rising average temperatures are expected to lead to an expansion of territories subject to constant droughts—from 2% today to 10% in 2050. Other things being equal, the consequences of such worsening conditions could be devastating. The IPCC sees the possibility of crop yields to go down dramatically in sub-Saharan Africa and in central and south Asia until mid-century, which are two of the most populous regions in the world.[25] If such projections hold true in their general tendency, the future movement of (parts of the) affected populations appears to be a likely scenario.[26] The worsening conditions will

---

[20] See Rebetez (2011, pp. 41–42).
[21] See Brown (2008, p. 17).
[22] See Rebetez (2011, p. 41).
[23] See Houghton (2015, pp. 178–200).
[24] See Brown (2008, p. 16).
[25] See IPCC (2015, p. 502).
[26] See Leighton (2011).

affect some of the most vulnerable countries and human communities. For people in such countries, a drier climate could leave them even more than before in a situation of abject poverty where migration figures as the only hope of alleviation, if not of survival.[27] There is evidence though that in some such cases even the (financial and physical) means to migrate may be lacking.[28]

Rising sea-levels: As a climatic driver that figures prominently not only in public debates but also meets with special interest in many scientific discussions on the climate-migration nexus, sea-level rise (SLR) should be considered here as a final and indeed special slow-onset migratory trigger. It will indeed play a special role in subsequent ethical discussions. Let me shortly consider the mechanisms behind sea-level and the projections on its impacts. The two reasons that cause this phenomenon to occur is first the fact that water expands in volume with rising temperatures (thermal expansion[29]) and second that there is water added to the oceans on account of the continuous ice loss, a process that was found to have accelerated over the last two decades.[30] According to current projections, the sea-level is likely to rise by at least one metre until the end of the twenty-first century.[31] It is beyond doubt that a rise of this magnitude will threaten human populations, and needless to say it will, depending on the region, do so long before it has risen by one metre.[32] Low-lying islands in the South Pacific will be affected much earlier—indeed they are affected already today.[33] Moreover, it is worth noting that rising sea-levels will tend to be accompanied by other detrimental processes like an increased proneness to floods. Such floods could be especially severe where the issue of SLR meets with poor coastal stability due to deteriorating conditions of mangroves and nearshore forests.[34] After all, among the climate triggers of climate change, sea-level rise is the most predictable and

---

[27] See Betts (2013).
[28] See Piguet *et al.* (2011, p. 7), Barnett and Webber (2010, pp. 40–41), and Hugo (2010, pp. 26–28).
[29] See Houghton (2015, p. 167).
[30] See Rebetez (2011, p. 42).
[31] See *ibid*.
[32] See Kelman (2008).
[33] See Boehm (2012) and for a discussion of concrete policy measures already taken regarding the relocation of communities on those atolls, see Campbell (2010).
[34] For a more complete account of the impacts of SLR on coastal areas see Houghton (2015, pp. 169–75).

clearest threat. There is high likelihood that in the long term it will cause permanent migration.[35]

5.1.3. Climate Change, Conflict, and Security

The final scenario to be considered here is migration in response to environmentally induced social tensions and conflicts.[36] The general mechanism behind the scenario in question is that climatic changes will in many cases lead to scarcities in fertile (grazing) land, forests, and, most importantly, water. Such reduced availability in resources could spark social tensions, unrest, and ultimately armed conflict, which could then trigger waves of refugees.[37] Especially in localities where climate change takes effect rather abruptly, where resources are already scarce, and where adaptive capacities are low (as they are in poor regions), carrying capacity may go down to the extent that a "Malthusian specter"[38] is raised. "The extreme case is that governments unable to protect their populations from climate change may collapse into 'environmentally failed states.'"[39] As one group of authors contends, such situations could culminate in "aggressive wars fought over food, water, and energy rather than ideology, religion, or national honor".[40] Presumably, such environmentally-induced conflicts will often be *perceived* by the affected populations as ethnic or religious

---

[35] See Piguet *et al.* (2011, p. 12) and Oliver-Smith (2011, pp. 165–70).

[36] For an insightful introduction to this nexus of climate change, migration, and security see Elliott (2010) and for comprehensive account of the ways in which climate change can end first in conflict and then in migration, see Smith and Vivekananda (2007). A further rough but concise overview of the (indirect) linkages between climate change and conflict is provided by Clark (2008) and for a short account of the alerting example of Darfur, where environmental stress can be shown to have contributed to social breakdown, see Edwards (2008).

[37] Kälin (2010, p. 86). Note, however, that the research on the link between climate change and conflict remains disputed. While a meta-analysis of 50 papers on that topic by Hsiang *et al.* (2013) concludes that the examined contributions consistently show support for the thesis of a link between climate change and conflict, Buhaug *et al.* (2014) question the results of this meta-analysis by pointing to alleged shortcomings with respect to the selection of the underlying sample and its analytical coherence. However, in a recent study of Schleussner *et al.* (2016) issued by the *Proceedings of the National Academy of Sciences of the United States*, the contributing authors find further evidence in global datasets that outbreak of armed conflict, at least in ethnically fractionalized countries, is enhanced through the occurrence of climate-related disasters.

[38] Gilman *et al.* (2011, p. 261).

[39] *Ibid.*

[40] *Ibid.*

conflicts, and this instance already raises the possibility that some forced climate migrants could perceive themselves or be perceived by "us" as war or Convention refugees.[41] In addition to tensions within the countries where the resource stress occurs, there is also the possibility of environmentally-induced conflict between countries, for example where two or more countries share a river basin or where a river flows between two or more countries and is (in response to water stress) dammed by the upstream country.[42] In all these scenarios, the refugees that are produced could themselves cause (further) resource depletion, localized violence, and eventually trigger migration in the places they themselves migrated to, be it international or inner-country migration.[43]

There is in public and academic debates a tendency to *frame* climate migration in terms of this nexus of climate change, migration, and conflict. Arguably, "the systematic inclusion of refugees, asylum seekers, and other categories of migrants on the research agendas of security scholars"[44] presents us with a moral issue *sui generis*. Therefore, before coming in the next chapter to a moral conceptualization of climate migration that departs from the just outlined migratory triggers, I consider it necessary to unfold in the following sub-chapter a moral case against that tendency to make climate migration a security issue or, to use the technical term that social scientists coined for referring to that tendency, to *securitize* climate migration.

5.2. The Case against Securitizing Climate Migrants

Is it acceptable to refer to climate migrants predominantly as a security threat? What effects will such securitization of climate migrants have? This is what it means to securitize an issue: "Issues become 'securitized', treated as security issues, through [...] speech-acts which do not simply describe an existing security situation, but bring it into being as a security situation by successfully representing it as such."[45] Similarly, Jane McAdam notes that "the 'lens' through which the phenomenon is viewed can dramatically change the way it is perceived and regulated."[46] If the lens through which academics, politicians, or citizens

---

[41] See Welzer (2010, p. 99).
[42] See Gilman *et al.* (2011, p. 261).
[43] Being displaced within their own country, they would in international law qualify as Internally Displaced Persons (IDPs) (see McAdam, 2012, pp. 250-52, and Kälin, 2010, p. 92). And for a short account of the protection gaps that remain for IDPs, see Koser (2008).
[44] Hammerstad (2014, p. 265).
[45] Williams (2003, p. 513).
[46] McAdam (2012, p. 15).

view the issue of climate migration is one of security, then fear rather than reason threatens to dictate our dealing with climate migrants.[47] My contention is therefore that by securitizing climate migrants, i.e. by degrading them to a security problem, one already and potentially *acts against their rights*, namely by failing to pay sufficient consideration to the moral specificity of their situation and thusly by failing to handle with sufficient care the primary question of what moral obligations we have toward them. Academics and policy makers are responsible for the frame they choose and for the real effects those frames have on real people.

I do not mean to suggest that any concern with security in the context of climate change and migration is ungrounded. Quite to the contrary, the previous section on the relation between climate change and conflict has put the finger on the well-understood mechanisms by way of which climate change *does* potentially heat up existing tensions and can lead to conflicts. To be sure, one should reckon with what Gilman *et al.* describe as the "extreme case that governments unable to protect their populations from climate change may collapse into 'environmentally failed states'" and that "[e]ven short of that limit case, states will produce large numbers of environmental refugees that will at minimum cause localized violence."[48] Such violence and conflicts may of course mean a security threat even to more distant countries. They could spread to neighbouring regions, and the migratory flows triggered by these conflicts may potentially include illiberal and radicalized people. And yet, there remains in the case of climate migration much *uncertainty* as to how much of climate-induced migratory flows will be cross-border, what groups will end up migrating, and of what quantity such flows will be in coming decades. So as Ingrid Boas, a leading scholar on the effects of securitizing climate migration, observes, it is precisely this uncertainty that "gives political actors much leeway to present and approach the issue of climate migration as they see fit [...]".[49] Too often, this leeway is used irresponsibly.

We need to recognize that "dominant forms of debate pertaining to climate refugees have been 'pegged out' as a security issue, when it could just as easily have been framed, for example, as a 'human rights' or as a 'development' issue."[50] And when political scientist Gregory

---

[47] For a trenchant critique of such alarmist depictions of climate migrants, see also Bettini (2013).
[48] Gilman *et al.* (2011, p. 261).
[49] Boas (2015, p. 9).
[50] Doyle and Chaturvedi (2011, p. 282).

White observes that "[c]limate change and security concerns have now fully merged in an increased elaboration of anxieties concerning climate-induced migration",[51] he suggests that the process of securitization is already well advanced. Indeed, one author even identifies the dominant perspective on climate migration to be one that draws on a militarist narrative in which "the climate refugee is constructed, at best, as a victim of a global polity with no human agency—a political entity *outside sovereignty*—or, even worse, as an environmental criminal or terrorist."[52] With a view to such advanced and potentially harmful processes of securitizing climate migration, I argue that securitization should be regarded as a moral problem in its own right.

> Securitizing CIM not only fails to solve the problem, but is also imprudent because it enhances security against a non-threat. [...] Treating it as a threat sets in motion a counterproductive, spiraling security dilemma that saps energies away from scientific analyses of the phenomenon, from the development of integrative policy solutions devoted to adaptation to climate change already underway, and from efforts to mitigate GHG emissions.[53]

Political actors can grasp frames and discourses as instruments for policy making, i.e. as channels to push one's case more effectively in public and political fora.[54] They know that it makes real differences with a view to policy outcomes whether an issue is finally framed in the language of human rights or in the language of national security. What this suggests is that what climate migrants get *from* us could very much depend on what we make *of* them. Worse still, to frame the issue of climate migration in terms of mass migration, geopolitics, and national security could well turn out to be a self-fulfilling prophecy. For where the focus is on the threat and the need to protect one's own country, the focus tends to be less on the need to mitigate and help (adapt), which in turn could lead to a larger actual number of climate migrants then perceived as a threat. From a universalistic moral perspective, the deeper problem with the security frame is thus that it is at its heart a particularistic frame: it looks at the interest of the few, and systematically neglects the interest of the many distant others. Even if not in denial of human rights, such a frame is at least disturbingly oblivious to concerns of justice—oblivious to the possibility that "we" helped bring into existence the very phenomenon that we now comfortably securitize.

---

[51] White (2011, p. 91).
[52] Doyle and Chaturvedi (2011, p. 279).
[53] White (2011, p. 125).
[54] For a thorough elaboration on this general point, see Gasper (2005).

It is problematic to declare climate migrants a security threat, and thus a risk. To construe them as a security risk depersonalizes the people behind the risks. It threatens to degrade the concern with human rights to a secondary one.[55] Securitization competes with a human rights discourse that asks what we owe to those climate-induced migrants. Indeed, the moral point of view urges us to avoid securitization processes as much as possible. After all, the very minimum we owe climate migrants is that we treat them respectfully, as bearers of human rights. We must not reduce them to a security risk — *even if* they were one.

---

[55] See Turton (2003, p. 10).

Six

Realizing the Conceptual Problems of Climate Migration

The previous account of the ways in which climatic changes can cause migration constituted a first step toward a moral conceptualization of climate migrants. Indeed, that previous chapter already made clear enough how difficult it is to speak of *the* (forced) climate migrant. Climate migrants constitute a most heterogeneous group which encompasses different sub-groups, like people leaving their homes in response to *sudden-onset* disasters, *slow-onset* environmental degradation, or even climate-induced conflicts. Moreover, corresponding to these various sub-groups, the concept of climate migration turned out to include different patterns of movement such as temporary and permanent migration, and internal and international migration. Now, in this chapter, a further layer of complexity will be added to the overall attempt of first conceptualizing and then discussing moral obligations toward climate migrants. It consists in two analytical challenges that the phenomenon of climate migration poses: (1) the difficulty of categorizing climate migrants and, more specifically, of putting them into *fixed* categories, and (2) the issues of causation. The later discussion of what we owe to climate migrants will build on this chapter's conceptualization.

6.1. The Fluidity of Categories

When the morality of immigration was discussed earlier in this work, it played a crucial role whether the migrating person is migrating "voluntarily" or whether she was forced to move. In other words, the degree of agency involved in a particular migrant's decision was central to attempts of categorizing and conceptualizing that instance of migration and that particular migrating person. However, as suggested before, the distinction between forced and voluntary migration is invariably difficult to draw. The degree of choice behind individual people's migration process is best represented by means of a continuum: while there may be clear instances of either voluntary or forced migration, there will always be in-between cases with a view to which it is by no means clear to what extent the affected persons are forced to move or not. If anything, the difference can then be captured as a gradual one, with one migrant moving more voluntarily than the other.

And yet, despite such difficulties, it turned out to be indispensable to have a binary distinction. In the face of scarce entrance places, a line must be drawn somewhere. In international law, protection norms are predicated on that distinction, which underlines the invariably binary nature of law.[1] Indeed, in previous parts of this work's moral analysis, a clear binary division into voluntary and forced migrants was also essential. It was essential for determining how strong the migrating person's claim to admission is. It determines whether she is considered a refugee or not.

Now, as I shall demonstrate in the following sections, the distinction between voluntary and forced migration gets even more complicated and blurry in the case of climate migration. It will turn out to be difficult to draw that distinction in practice. Nonetheless, I will in a first step make use of that distinction in order to propose how—at least *in theory*—one could meaningfully distinguish between three different groups of climate migrants. In subsequent steps, I will then raise awareness of the difficulties involved in drawing on those categories *in practice* (a recognition which will then, later in this work, have far-reaching implications on the question of what we owe to climate migrants in general).

6.1.1. Categories of Climate Migrants, in Theory

The International Organization for Migration (IOM) employs the following working definition for "climate change migrants":

> Persons or groups of persons who, for compelling reasons of sudden or progressive changes in the environment as a result of climate change that adversely affect their lives or living conditions are obliged to leave their habitual homes, or choose to do so, either temporarily or permanently, and who move either within their own country or abroad.[2]

This definition covers a wide range of sub-types of climate migrants. It subsumes under one single category a diverse group of climate migrants, of which some are forced to move while others are not. This

---

[1] Kälin (2010, p. 95) and Zetter (2010, p. 140).
[2] International Organization for Migration (2008, p. 31).

clearly compromises the analytical helpfulness of the IOM's definition. Moreover, the IOM's definition is not sensitive to the circumstance that particular instances of climate migration—especially when they are due to slow-onset processes—are determined by a *timing factor*: whether a migrant is forced to move or not often depends only on *when* the person leaves the affected area. This *temporal dimension* should not remain unconsidered. Instead, I will here draw on it centrally when I develop a more fine-grained categorization of climate migrants and especially in subsequent discussions of our obligations toward climate migrants it will play a central role. To begin with, I propose to distinguish between three (theoretically plausible) types of climate migrants:[3]

(1) There is a first category of those who move considerably *before* environmental conditions become insupportable (for themselves and/or for their households). Their movement can be considered a form of adaptation either to already perceptible *slow-onset* climatic changes or, as will be substantiated later, to the perceived risk of being hit by the effects of a threatening *sudden-onset* disaster. As concerns this first strategy of *migration as adaptation* there is clearly some relevant degree of choice involved, with migration figuring as a possible option which individuals, households, or communities that are faced with actual or perceived climate threats can draw on in order to reduce the risks associated with adverse climatic conditions.[4] Both the actual deterioration of the environmental conditions and the *perceived* risk of suffering sudden loss and damage can induce (i.e. motivate) affected people to move,[5] so that I henceforth refer to this kind of climate migrant as *climate-induced migrants*.

---

[3] The following three categories are in important respects informed by Hugo (2012) and Renaud et al. (2007). Renaud et al. (2007) distinguish between "environmentally motivated" migrants, "environmentally forced" migrants, and "environmental refugees", and Hugo (2012) largely adopts and refines this categorization. However, my own categorization diverges terminologically and substantively from these two and other attempts in the literature of distinguishing between climate migrants. With my explicit focus on risk perception as one driving force of migration (informing the first and the second category) my categorization diverges from Renaud et al. substantively. And I diverge from them terminologically in so far as I speak of the *climate* rather than the *environment* (which is a broader concept) as the driving force of migration. The difficulty of denoting a certain environmental change as climate change induced is addressed separately below, when I discuss the issues of causation.
[4] See McLeman (2015, p. 63).
[5] See Hugo (2010, p. 13).

(2) Then there are those who migrate only when, in consequence either of continuous *slow-onset* environmental degradations or due to a (by now) unacceptably high *risk* of becoming the victim of a sudden-onset disaster, living conditions have become unendurable to the point of eventually *forcing* the affected people to leave their place of living. Although forced to move in the relevant sense of dispositional freedom, they may retain at least some degree of choice regarding the timing of their movement.[6] I will refer to this group of climate migrants as *forced climate migrants*.

(3) Finally there is a group of climate migrants who, in response to materialized *sudden-onset* environmental disasters, are forced to move or rather: *displaced* by the catastrophic event. For the affected, there tends to be no relevant degree of choice left; they cannot even make a decision on the timing of their movement.[7] I henceforth refer to them as *climate displaced persons* (CDPs).[8]

Note that this categorization cuts across the distinction between *sudden-onset* disasters and *slow-onset* environmental processes.[9] The victims of sudden-onset disasters figure not only in the third category of "climate displaced persons". Rather, sudden-onset disasters also play into the first and the second category of *climate-induced migrants* and *forced climate migrants*. This integration of sudden-onset disasters into the first two categories (i.e. categories which are, presumably, more naturally associated with slow-onset processes than with sudden-onset disasters) becomes possible only by taking seriously the *perception* of climate *risks* for the "decision" to move. Indeed, it is *the perception of climate risk* that often leads to an increasing rate of emigration from an area affected by such perceived risk.[10] Sudden-onset disasters induce or force people to move even before they hit; they do so in their pre-materialized form of real or perceived environmental risk. As a matter of fact, and with

---

[6] See *ibid*.
[7] See Hugo (2012, p. 13).
[8] This term of "climate displaced persons" is also used by Lyster (2015, p. 430). Note, however, that she uses this term in a much broader sense, that is, to cover a much broader range of cases. In fact (and indeed little helpfully), it serves in her analysis as an umbrella term for all climate migrants.
[9] What is not captured by this categorization is the additional complexity that arises from the earlier noted tendency that slow-onset processes tend to lead to permanent movement while sudden-onset triggers are associated rather with temporary movement. After all, the distinction between temporary and permanent migration will turn out to be a rather secondary concern for the question of what we owe to climate migrants.
[10] See Hugo (2010, p. 15).

regard to such risks of sudden disasters, governments might designate certain areas or regions as "high-risk zones too dangerous for human habitation"[11] and evacuate them on such grounds. The recognition of the importance of climate risks will figure centrally in following attempts to spell out what we owe to climate migrants.

6.1.2. Voluntary vs. Forced Migration: The Impracticality of a Central Dichotomy

After this proposal of how one could in theory make a meaningful distinction between different groups of climate migrants, I will in this section point out why this categorization is nonetheless practically problematic, because it (still) relies on a doubtful distinction between voluntary and forced movement.

The decisive observation on which I will draw and which is not new at this stage is that the environmental deterioration that is caused by slow-onset climate-induced processes is often *incremental* in nature. As I will argue in the following sections, the gradual nature of this migratory trigger makes it "less easy to argue that climate change 'compels' or 'forces' displacement [...]".[12] Most people will not stay until the water is up to their knees. They will move (long) before. It therefore remains unclear when they should be conceptualized as *choosing* to migrate or as being *forced* to do so. The same could also be assumed of people living in regions where hazardous hydro-meteorological events become *continuously more frequent*. Many may leave well in advance of that moment in which they can be said to have no choice but to move. Such reasoning already informed the above mentioned "timing factor". In the remainder of this section I will provide an account of the fundamental importance of this timing factor for conceptualizing climate migration. And based on this account, I will then propose a more holistic way of understanding climate migration, namely one that begins temporally and geographically in the home country, i.e. in situ. There, the affected people still figure as *potential climate migrants*. This more holistic perspective will complement the above categorization.[13]

What the timing factor discloses is that the line between the first (i.e. *climate-induced migrants*) and the second (i.e. *forced climate migrants*) of the above categories is blurry: at least in some cases — depending on the

---

[11] Kälin (2010, p. 91).
[12] Zetter (2010, p. 140).
[13] See Hugo (2010, pp. 12–20). Some of my following proposals are inspired by Hugo's valuable conceptual work on climate migration. In so far as Hugo does not take a moral perspective though, I go beyond his account in relevant respects.

ways in which actual environmental changes proceed or in which risk perceptions alter—the difference between these first two groups appears to be only a *temporal* one. If only the *climate-induced migrant* had waited long enough she would have equally been forced to move at some point, thus ending up as a *forced climate migrant*, and any *forced climate migrant* could have been described as a *climate-induced migrant* if she had left at an earlier stage, i.e. before life became unbearable due to continuous environmental deteriorations or due to an increased perception of climate related risks.

This fluidity between the first and the second form of climate-related migration arguably has important morally-normative consequences for the question of what we owe the (potentially) affected. After all, it appears to suggest that (at least in many cases) the degree of force that is involved in a particular climate migrant's movement should not determine what we owe to that particular migrant. Is her claim toward us less powerful only because she was wise enough to leave before life became unbearable? Should the *climate-induced migrant's* claim be rejected, advising her instead to come back later when life in her country has become sufficiently unbearable so that she finally fits into our category of *forced climate migrant* as opposed to the "mere" *climate-induced migrant* she is now? This rhetorical question must be negated. The alternative would be to hold that *if* we have strong moral obligations toward *forced climate migrants* then due to the described fluidity of categories and due to the *arbitrariness* of the timing factor it seems we have comparably strong moral obligations toward the first group of *climate-induced migrants*.

To the extent that the line between the first category and the second category of climate migrants is (often) fluid and will (at least in some cases, especially where slow-onset processes are concerned) depend only on a timing factor, we should consider including *climate-induced migrants* under the heading of "being forced": for they are often at least *potentially* forced and—on pain of repeating myself—it would be dubious to accept moral obligations only toward those who endured the worsening environmental conditions and the higher perception of risk long enough. To ask whether someone, at a particular moment of her flight, is forced to move or not would lead to a rather arbitrary determination of who is morally entitled to protection. As regards slow-onset degradation, the degree to which a migrant is actually forced tends to depend largely on when (i.e. at what specific moment) in the overall process of environmental deterioration the situation is assessed (by us). As an important observation then, climate migrants— with the time passing—often move downward (on the spectrum of categories) toward the category of *forced migration* or (waiting for the risk of sudden events to materialize) toward *climate displacement*.

This observation can also be made meaningful in terms of the substances of the affected people's basic rights: they move from "possession of the substances of basic right" (i.e. voluntary movement) to "deprivation of the substances of basic rights" (i.e. forced movement). But this last observation is true only in the sense of occurrent freedom as introduced earlier. In the sense of dispositional freedom it would be an inadequate description of reality. By taking dispositional freedom as the relevant standard it becomes visible that people who appear to fall into the first category of climate-induced migrants may often not (fully) enjoy basic rights either. They may have the substances of basic rights in the *occurrent* sense, but living in constant awareness that they may (or depending on the case: that they *will*) have to leave sooner or later, it is at least questionable to speak of these migrants as (truly) enjoying basic rights and therewith as moving voluntarily.

Therefore, I propose to understand climate migration in a more dynamic and holistic way. The perspective taken when the claim of climate migrants is assessed must be a *diachronic* one. It must be sensitive to the described fact that the urgency of the climate migrant's claim is, often predictably, subject to changes over time and that by tendency the claim will become more rather than less urgent. One should thus be wary of putting climate migrants into rigid categories that can account neither for the dynamics of climate-induced slow-onset environmental degradation nor for the possibility that exposure to the risks of sudden-onset disasters also increases with the proceeding of climate change. The question of what we morally owe to climate migrants must not depend on the arbitrariness of the moment in which the affected individual presents her claims to us (although, as I argue in the next section, this moment is of course relevant in so far as it determines the possible policy measures by way of which that affected individual can actually be assisted).

After all then, the here discussed limited reasonableness of putting climate migrants into the fixed categories of *climate-induced migrants* and *forced climate migrants* has a radical implication: in a morally relevant sense, the claim of the *climate-induced migrant* converges toward that of the *forced climate migrant* (or of the *climate displaced person*). In so far as either category represents just a different stage of a process (or continuum) that *tends* to lead from the first toward the second (or third) category without clear lines dividing the two, the substances and freedoms which the affected individuals from either category have at stake are—in the truest sense of the word—

"potentially" (and sometimes predictably) the same. With regard to the *right to exclude* as formulated in the third chapter, this insight has potentially far-reaching implications. The decisive consideration is that many climate migrants, even those who in the particular moment at which they reach a potential host country would still figure as voluntary *climate-induced migrants* because they emigrated considerably *before* their region became uninhabitable, make the very powerful claim that their basic rights are insecure in the medium or long-term and that they thusly (similar to refugees) do not enjoy dispositional freedom in their home country. The exclusion of a potentially large number of migrants who can bring forward such a powerful moral claim to admission would appear to be morally problematic indeed.

It turns out that the here made conceptual work has the potential to challenge the way in which (parts of public and academic) normative debate on immigration policy are currently structured. Such debates are indeed organized around the distinction between voluntary migration and forced migration, and correspondingly around the distinction between normal migrants and refugees. I followed this example in my above dealing with the morality of immigration and there can be no doubt that this distinction is normally instructive and — against the background of scarce entrance places — practically necessary. And yet, to the extent that this distinction becomes impractical for the case of climate migration (which the above conceptual analysis clearly suggests), the morality of immigration in its current form also requires some re-assessment. This would be especially true if and to the extent that in the future the largest group of worldwide migrants were indeed climate migrants.

6.1.3. Potential Climate Migrants: The Neglected but Decisive Category

Departing from the previous section's finding that climate migrants tend to undergo a process through the categories, I will in this section propose a new way of thinking about climate migrants which grasps the *whole migratory process*. This process invariably begins in the affected people's home country. There, they are not yet climate migrants, but only figure as *potential climate migrants*. Indeed, it is suggested in this section that the *potential climate migrant* is the *decisive* (new) category from which all normative reasoning on what is owed to climate migrants should depart.

In the following, I will present three scenarios, each of which literally departs from the affected person's perspective in the home country. In the first scenario the affected person can remain in situ, in the second she becomes a *climate-induced migrant*, and in the third she ends up as a *forced climate migrant* (or as a *climate displaced person*). I use this method of sketching scenarios in order to (1) highlight the decisive recognition that the migration process invariably begins in the home country and (2) that it is there — in the home country — that the course is set either for staying, leaving, or fleeing. In other words, whether the *potential climate migrant* stays, leaves, or flees depends on what happens when the potentially affected person is still in her come country; it depends not only on how the environment changes, but just as much on the adaptive capacities the affected individual (or community) has; it depends on the means she has for helping herself and on the means she is provided by third parties. And this is where the question of what we owe to climate migrants would come in.

After all, the new focus on the climate migrant when still in her home country comes with the important conceptual advantage that it sidesteps the above observed impracticality of the distinction between voluntary and forced movement. When still in situ, she does not move in either way, but is primarily viewed as a bearer of certain rights (for example a right to stay) whose options depend on how her environment changes, on her own decisions, and on what others do about her situation and environment. This is the most promising starting point for my later reasoning on obligations toward climate migrants. I will now turn to the three scenarios.

1. The *potential climate migrant*'s vulnerability to the increasingly perceptible effects and/or threats of climate change is reduced through policy measures which allow the *potential climate migrant* to adapt in situ to the detrimental environmental changes or to the increasing perception of climate risk (i.e. the risk of becoming the victims of sudden-onset disasters).

    In this first scenario, the *potential climate migrant* could stay; she then becomes neither a *climate-induced migrant* who moves in order to adapt to the environmental changes, nor does she become a *forced climate migrant*.

    Now, concerning the first scenario, what the affected person would claim in that early stage is, couched in the language of rights, the *right to stay*.[14] This right would, from the affected person's perspective and depending on her particular context,

---

[14] As Jane McAdam notes with regard to this emphasis on the right to stay: "This shift in approach does not downplay the effects of climate change, but rather reasserts the population's right to remain and focus efforts on in situ adaptation strategies" (McAdam, 2012, p. 35).

appear to be at least as valuable as any other form of adaptation.[15] An important practical reason why responsible countries should first and foremost concentrate on the effectuation of this right to stay is (beside the possibility that staying is the preference of the migrant herself) that it constitutes a form of assistance which potentially helps *all* people in the affected country in a sustainable manner, including the poorest members of the affected societies (or regions). Even if other countries allowed them to immigrate, they and their households often lack the financial means to seize that opportunity. It is in this sense that only the fulfilment of the right to stay truly reaches to the needs of the poorest.[16]

2. The *potential climate migrant's* vulnerability to the increasingly perceptible effects and/or threats of climate change is *not* reduced through policy measures that allow the *potential climate migrant* to adapt to the changes and threats in situ.

   In the absence of such in situ adaptation measures offered to the *potential climate migrant, (cross-border) migration as a form of adaptation* may at some stage appear to her as a suitable option. Whether and at what point in time she will indeed take this adaptive strategy will depend on various factors like her own preferences (timing factor), the financial situation of her household, the possibilities to migrate within her home country, and finally the channels and possibilities for international migration. In this second scenario, she would become a *climate-induced migrant*.

   What the *potential climate migrant* could claim in that second scenario is the right to adapt by immigrating into another country. To the extent that a right to stay is not or cannot be granted, and in light of continuous environmental degradation and/or the risks of sudden-onset loss and damage that she perceives, enjoying the right to move across borders and seize opportunities in other countries appears to be the second-best way of protecting the victim's right to long-term freedom and basic well-being. Indeed, it will in particular cases be the most promising (if not the single) way of regaining the dispositional freedom which, in the face of the constant threat of climate change, she no longer enjoys. Such migration to another country is often not only an adaptation strategy for the affected individual, but also for her family and even for the

---

[15] For the view that the right to stay and thus to adapt in situ is at least as important as the right to leave and thus to adapt by migrating, see Barnett and Webber (2010, p. 50). As geographers, they do not try to ground this view morally though.

[16] See Hugo (2010, p. 26) and Barnett and Webber (2010, p. 40).

affected region as a whole. Empirical studies found that in rural Asia and the Pacific, families try to cope with the effects of climate change by deploying individual family members to other labour markets inside their own country and internationally.[17] "While the volume of resources sent home by poor migrants may be small, the relative contribution to household incomes and capital is large and so significantly increases adaptive capacity."[18] This environmentally-induced labour migration thus promises to have positive effects not only for the family but for the affected country (or region) as a whole. When some of the people from those countries leave, then this "out-migration can have the effect of reducing the pressure on resources in the origin, but also through remittances sent back can reduce the vulnerability of [all] those left behind and increase community resistance."[19] What such broader positive effects suggest is this: if we are responsible for those affected people and if we are not willing to, or cannot, help them adapt in situ, then we should not obstruct their efforts of helping themselves by moving internationally.

3. The *potential climate migrant*'s vulnerability to the increasingly perceptible effects and/or threats of climate change is not reduced through policy measures that allow her to adapt in situ, nor can she draw on international migration as an adaptive strategy (because she lacks the means, the willingness, or the channels to do so).

This means she does not become a *climate-induced migrant*. In the course of time and to the extent that either the (perceived) risk of being hit by sudden-onset catastrophes gets unacceptably high or that in consequence of long proceeding slow-onset processes the actual environmental conditions no longer provide for only the minimal means needed to make a decent living, she will—in the absence of other forms of protection and assistance—be *forced* to leave her home. She would then, eventually, be a *forced climate migrant* or, if a sudden-onset disaster occurs, a *climate displaced person*.

Concerning the third scenario, it goes without saying that the right to leave as a *forced climate migrant* and the (hopefully granted) corresponding right to get shelter in another country constitutes only the third-best option for assisting the affected people. However, the moral rationale for providing that assistance is the least

---

[17] See Hugo (2010, p. 14).
[18] Barnett and Webber (2010, p. 44).
[19] Hugo (2010, p. 15).

disputable one. Here, assistance is granted to people already fully deprived of the substances of basic rights or at least to people who are at high risk of being deprived soon. But serious doubts were already cast on the meaningfulness of this third-scenario form of assistance: as concerns slow-onset degradation, the potential victims will at some stage require assistance at any rate—so why have them go through the painful course of that process rather than assist them by granting either of the two earlier forms of assistance (i.e. to stay or migrate as adaptation)? It is intuitively appealing to hold that there is a responsibility to protect the potential victims from becoming full-blown victims in the first place. Once they are full-blown victims, they would then need to be rescued in any way.

Note that the here proposed special concern with *potential climate migrants* rather than with actual climate migrants responds to an issue that has already been discussed before: why should the destination country help those migrants and refugees who make it to their shores rather than all the other people who have not reached that destination country but remain on their dangerous journeys, in far-away refugee camps, or in their own countries although they are not safe there?[20] The focus on the *potential (climate) migrants* is a response to this problem and it remains to be seen in subsequent chapters how this focus on the people in situ can be converted into more concrete *political* proposals.

6.2. The Issues of Causation

In fact, many hydro-meteorological disasters triggering displacement would occur regardless of climate change, and even where they are linked to climate change, such causality is difficult, if not impossible, to prove in the specific case.[21]

A major challenge faced when attempting to spell out what is morally owed to climate-induced migrants is that the causal relationship between climate change and migration is difficult to establish. I will refer to this challenge as the *epistemic issue of causation*. For specific events, knowledge on the exact causal linkages between climate change and migration is limited. There will in many cases remain uncertainty on the particular causal role of climate change in bringing about specific weather events which in turn (together with other causal factors) triggered migration. In fact, this difficulty of establishing causation for specific events is one that pervades climate sciences more

---

[20] See Gibney (2004, p. 10).
[21] Kälin (2010, p. 84).

generally: despite some strong evidence that there is an increase in climate extremes due to anthropogenic climate change, the attribution of single weather extremes to anthropogenic climate change remains a challenge.[22] As regards the few ethical accounts on climate migration, a systematic debate of how to deal with the difficulty of establishing a causal link between anthropogenic climate change and the specific instance of migration is widely neglected.[23]

6.2.1. The Epistemic Issue of Causation and Attribution

Assume a sudden-onset disaster like a destructive hurricane. How can one know this Hurricane would not have taken place in the absence of human-induced climate change? In fact, many events might have materialized (or proceeded) regardless of anthropogenic global warming, although presumably at a different point in time and with varying severity. This is true both for sudden-onset hydro-meteorological disasters and for slow-onset processes such as land degradation.[24] Often, it may "only" be the case that climate change intensifies such processes, but even for this verdict there may remain some uncertainty.[25] It is against the background of such uncertainty that Zetter advises to be "cautious in attributing every negative environmental condition to climate change".[26]

However, even if single events cannot be attributed to climate change, the accumulation of (an increasing number of) such single destructive events can—on probabilistic terms and then potentially with a high degree of certainty—be attributed to the changing global climate.[27] After all, the empirical problem of identifying climate change as the cause of a particular migration-triggering environmental event or process grounds an important conceptual problem: it appears to be dubious whether one can speak in meaningful ways of "climate" migrants (rather than of "environmental" migrants), when it is not clear whether and to what extent it really was the *changing climate* that caused the movement of a *particular* group of people.

---

[22] See IPCC (2014a, pp. 234–36).
[23] A salient example of this shortcoming is Nawrotzki (2014). While emphasizing that industrialized countries are morally responsible to assist climate migrants precisely because there is a putative causal link, he simply ignores the decisive difficulty that this putative causal link is often impossible to establish for the specific instance of climate migration.
[24] See Kälin (2010, p. 85) and Zetter (2010, p. 138).
[25] See McAdam (2012, p. 93).
[26] Zetter (2010, p. 138).
[27] See Jacobeit and Methman (2007, p. 12).

Moreover, even *if* it could be established that it was emissions that caused a certain environmental damage, it would be impossible to *attribute* this damage to a particular polluting country, because greenhouse gases mix in the atmosphere. I will refer to this aspect of the epistemic *issue of causation* as the *problem of attribution*.[28] Some authors argue that, as it is not possible to establish some particular parties' emissions of greenhouse gases as the cause for environmental harms and for the ensuing migration (and displacement) of people in affected countries, it is difficult to apply a corrective justice frame and thus to hold putative causing countries responsible for their actions.[29] Legal theorist Jane McAdam notes in this respect that the "global nature of climate change, and the fact that alleged harms may be 'geographically and temporally divorced from the adverse consequences', mean that it will be difficult for an applicant to show a sufficiently direct or specific relationship such as to establish a duty of care."[30] To what extent this difficulty of proving a specific causal relationship constitutes a problem for the applicant (as McAdam suggests) remains to be discussed below. I will defend the thesis that climate migrants can sidestep the issues of causation and thusly *can* ground their claims (also) on the principle of corrective justice.

It stands to reason that short of being a "mere" conceptual problem, the *epistemic issue* of causation lies at the root of far-reaching moral and ultimately political questions about responsibility for the victims of climate migration. In other words, behind the conceptual difficulty identified here lies the real-world moral and political challenge that someone must be made responsible for those people who are harmed (i.e. displaced) by human-induced events and processes. It appears to be intuitively unacceptable that polluting countries, simply by pointing to the absence of a clear evidentiary link between their particular actions (i.e. emissions) and other people's specific instance of displacement, can abdicate their responsibility. In international law as of today there is indeed a normative gap that would allow countries to eschew their responsibility for the fate of climate migrants.[31] However, the conceptual difficulty associated with the here outlined *epistemic* issues of causation and attribution can only partly explain what is so challenging about realizing effective legal protection for the rights of people displaced by climate change. There is another aspect to the issues of causation that renders it difficult for affected people to raise claims to

---

[28] See Zetter (2010, pp. 131–50).
[29] See Wyman (2013, pp. 193–95).
[30] McAdam (2012, p. 93).
[31] See Zetter (2010, p. 139).

compensation against presumably responsible parties: the *issue of multi-causality*, which will be outlined in the following second section.

6.2.2. The Analytical Issue of Multi-Causality

Even where a causal link can be established—or rather, where on probabilistic terms one may assume that there is such a link—there emanates a further problem from the fact that the climate is in most cases only *one* factor that explains migration. Indeed, there is normally a wide range of further variables that play a role. Generally speaking, such variables can be political, social, demographic, or economic in nature.[32] To be more specific, "[p]olitical, structural, and idiosyncratic factors, such as ethnicity, family size, productivity in potential destination zones, the agrarian state of rural areas, schooling opportunities, social networks (in destinations), and the availability of food and water in other areas"[33] all have an impact on the decision of individuals to move. Climatic factors then figure as just one stressor that interacts with, adds to, or multiplies these "pre-existing" stressors. It is thus in *combination* that they produce migration.[34] What this suggests is that, in most instances of migration, the decision to migrate is characterized by multi-causality:

> Discussing multi-causality [...] implies acknowledging the non-direct relationship between climate change and migration, and the factors that mediate between the two. Climate change is clearly a complex environmental process that does not have uniform consequences everywhere; and societies have always had to adapt to changing environmental contexts—a multifaceted process of technological, organizational institutional, socio-economic, and cultural nature that is likely to be just as complex as climate change itself. The number of variables is therefore important, leading to high uncertainty and local variability.[35]

The crucial practical implication of this observation is that it will in most cases not be "possible to identify a group of people who migrate *only* because of environmental variables".[36] Conceptually, this recognition entails serious problems. It further calls into doubt the adequacy of speaking of "climate migrants". If the climate is only one among

---

[32] See Hugo (2010, p. 10).
[33] McAdam (2012, p. 21).
[34] It is with a view to such difficulties of multi-causality that some authors regard it as problematic to make any serious estimates on the number of people to be expected to move as climate-induced migrants in the future. See Foresight (2011).
[35] Piguet *et al.* (2011, p. 13).
[36] *Ibid.*, p. 18.

other social, political, and economic factors, then why reduce the migration that is triggered by them to "environmental" or "climate" migration? Indeed, as one author sees it, the term "environmental refugee" is inadequate in so far as it is "simplistic, one-sided and misleading. It implies a mono-causality which very rarely exists in practice".[37]

From a moral standpoint, a central problem with multi-causality becomes clear when we reconsider Rawls' notion (as introduced in chapter 2) that—for reasons of incentive structure—individual states should take care of their own assets and should (at least to some extent) themselves bear the consequences of mismanaging their own resources.[38] If due to the difficulties related with the *issue of multi-causality* it is often not possible to say whether and to what extent it is the individual country itself that is responsible for a given suffered loss (in the sense that domestic political, economic, and social factors primarily account for the loss) or whether it is mainly third parties that are responsible for that loss (that is, because it is primarily the climatic factor that accounts for the loss), then Rawls' rationale would clearly be undermined. The distribution of responsibility and therewith the discussion of who is responsible to what extent for (potential) climate migrants becomes a most complex endeavour in the face of the issue of multi-causality.

In terms of migration policy, an immediate problem that the awareness of the multi-causal nature of most migration processes leaves us with is the difficulty of distinguishing in a meaningful and practical way between climate migration and other types of migration like poverty (or economic) migration or political migration (encompassing political or Convention refugees). I will now, very briefly, raise and assess a series of different approaches that one could draw on in order to deal with this difficulty of grasping the overall category of climate migrants. Note, however, that the following list and short assessment of possible approaches to the question of how to mark off climate migrants from non-climate migrants does not present a "solution" to the issue of multi-causality. I will only point out who can meaningfully be labelled a climate migrant in the face of the issue of multi-causality, but I will say nothing on the question of how one could specify our obligations toward climate migrants in the face of the issue of multi-causality. This latter issue is dealt with only in chapter 7 when I debate the "insurance approach".

---

[37] Castles (2002, p. 5).
[38] See Rawls (1999, pp. 38–39).

Now, a first approach would be to say that if only the climate plays *some* causal role, it makes sense to call the affected people "climate migrants".[39] However, to have a category of "climate migrants" that effectively includes most of the world's migrants (in so far as the environment is *somehow* involved in all of the world's migration) would seem to render that category analytically useless or, at least, politically impractical.

A second approach to this conceptual problem would be to label as "climate migrants" only those people for whom the climate constituted the *primary* cause behind the individual decision to migrate. But needless to say, any attempt to extract the climate as the *primary* reason would confront us with the same epistemic and analytical problems that were discussed above. Indeed, "identifying the 'primary' cause of migration is probably impossible, as all causes may mutually reinforce each other."[40] It is against the backdrop of this difficulty of isolating the environment as the primary cause that Richard Black—one of the earlier and most widely cited contributors to the scientific debate on climate (or environmental) migration, calls into question the meaningfulness of defining a category of environmental or climate refugees: although Black does not deny the real possibility and importance of environmental factors as a driving force behind migratory processes and displacement, he regards "their conceptualization as a primary cause of forced displacement [as] unhelpful and unsound intellectually, and unnecessary in practical terms".[41] A defence of this claim leads Black to the conclusion that, in a world where migratory decisions are usually the result of several interacting factors, a meaningful definition of environmental migration "does not seem very likely"[42] and that—in so far as a properly defined category does not exist—it will not be "easy to say that this category of people is increasing".[43] He then concludes that it will not be easy to deal with an inexistent category in institutional terms.

This reasoning of Black on the conceptual difficulties of climate migration—and especially of the issue of multi-causality—is not persuasive. Even if it is not easy to come to terms with a precise and unambiguous definition of (forced) climate migrants, it is quite another thing to say that this difficulty will hinder us from tackling the phenomenon of climate migration institutionally or even from assess-

---

[39] See Jacobeit and Methman (2007, p. 14).
[40] Piguet *et al.* (2011, p. 13).
[41] Black (2001, p. 1).
[42] *Ibid.*, p. 14.
[43] *Ibid.*

ing whether the number of affected people is increasing. Black appears to move from definition to reality, so when a definition cannot be found, the reality cannot be dealt with either. This is a bizarre way of dealing with a generally understood issue like climate-induced migration. For after all, we understand the general mechanics behind climate migration, despite the mentioned difficulties regarding the specific instances of climate migration. As I will contend here and in subsequent parts, there *are* practical strategies for coming to terms with the phenomenon of climate migration, blurry and multi-factorial though such migratory processes may be. A first step toward these strategies lies in the recognition that, contrary to Black, the (almost impossible) identification of the climate or the environment as the "primary cause" of a migratory process is not the only — let alone the most meaningful — way of marking off climate migrants from other groups of migrants.

As a third approach to the issue of multi-causality, it would also be conceivable to speak of "climate migration" *only* in those cases where the climate is the actual and *final* trigger of migration, i.e. where the climate is the *intervening variable* among other context-factors that remain in the background and would not by themselves have triggered migration. In other words, one could look for those cases in which the climate plays the role of the last straw that breaks the camel's back. When people are poor (background variable) and then have to leave their homes because of rising sea-levels (intervening variable), then, as some authors suggest, it would be wrong to call these people economic migrants or poverty migrants.[44]

To see why this focus on the climate as an intervening variable is inappropriate as well, consider the following imaginative example of a Guatemalan peasant. She lives in a region that through climatic changes has become more arid than it used to be. The peasant is poor, but she gets along, so that she does not have to migrate. She may be on the verge of having to do so, but fortunately the climatic changes that had previously deteriorated her livelihood have now ceased to make the situation continuously worse. She can stay. As this case is constructed, climate change is just a background variable. So let us now introduce an economic intervening variable: the economic conditions worsen, and this occurs either because the peasant mismanages her assets herself *or* because of governmental failures in economic policy. Recall that she was already on the brink of displacement, and now it is these economic changes that function as the last straw that breaks her back. She has to leave her place of living, let us assume internationally,

---

[44] For such an approach, see Jacobeit and Methman (2007, p. 14).

as Guatemala cannot offer her any alternative livelihood. By the above proposal, she is not a climate migrant, but an "economic" or "poverty" migrant. This assessment appears to be inappropriate. It neglects the climate only because it does not figure as the actual trigger but played a causal role earlier in the overall process that finally lead to the migration of the Guatemalan peasant. The recognition of this arbitrariness allows me to reject that third approach and leads me to a fourth way of dealing with the issue of multi-causality.

From the perspective of the victim, it is irrelevant which variable takes its destructive effect at what stage of that overall process which ultimately culminates in her migration. This is what the example of the Guatemalan peasant meant to make clear. For her, what matters is that she *likely* (or at least potentially) would not have had to migrate if there were no climatic changes (or the respective other factors). This suggests the fourth and final approach that I will embrace here: that a migrant may be considered a climate migrant if climate change is a *sine qua non* condition of her migration or displacement. As with all other approaches just discussed, the epistemic issue of causation appears to remain an obstacle in the way of applying it in practice, because it appears to remain difficult in the specific case to establish the climate as a *sine qua non condition*. But conceptually and morally, this approach to the issue of multi-causality is promising because, as the example of the Guatemalan peasant made clear, it is sensitive to the perspective of the affected migrant. For her it is a decisive recognition that she would not have been induced or forced to migrate but for the occurrent climatic change.[45]

---

[45] Note that this focus on *sine qua non* conditions would in principle allow for the circumstance that a particular migrant can point to more than one factor that in his particular case plays the role of a *sine qua non* condition, for example both the climate and the economy. But for the present attempt of defining the group of climate migrants, it hardly presents a serious problem that some of the migrants who are considered climate migrants might *also* count as economic migrants or even war refugees. It wouldn't be worrisome if, assuming there will one day be protection for a separate legal category of "forced climate migrants" (or whatever the term may then be), some migrants qualified both as Convention refugees and as "forced climate migrants".

Seven

Responsibilities toward Climate Migrants

This chapter aims at *overcoming* the analytical and conceptual difficulties that were pointed out in the previous chapter and that so far remain unsolved. The present chapter consists of two parts (or sub-chapters). In the first part, I will propose a strategy for *sidestepping the* issues of causation. By developing a *sequence* of nested argumentative steps, I will contend that under certain conditions at least some emitting countries have a corrective justice-based responsibility to assist climate migrants. That argumentative sequence will combine for a new perspective on the issues of causation and it will respond to most of the previously discussed reservations against applying corrective justice to the case of climate migrants.

In the second part of this chapter, which directly builds on the findings of the first part, it will be discussed in what more concrete ways our (corrective justice-based) responsibility toward climate migrants could be discharged. In that final account, I will call for responsible parties to direct their efforts at taking certain preventive or (as I will call them) *ex-ante* measures.

To be clear, what I propose to be our moral obligations toward climate migrants will not be grounded in the principle of corrective justice alone. As I shall argue toward the end of this chapter, there are various normative sources that may complement my case for assisting climate migrants, especially the general thrust of the human rights perspective which grounds a duty to assist needy people simply because one is capable of assisting them. What justifies my continuous concentration on corrective justice though is not only that it is intuitively appealing and of extraordinary normative power, but primarily that in the particular case of climate migration it is much more contested than the concern with basic human rights. Specific climate migrants might blame emitting countries and demand compensation, and the emitting countries might reject that specific claim as insufficiently backed by evidence. In this heated context, sober moral reasoning is needed. If it can be shown that and how climate migrants can convincingly draw (also) on corrective justice for making their claims, then this would finally make sense of the intuition that the kind of moral demands that climate migrants make on us are indeed different from the demands other (non-climate) migrants raise. The difference would lie in the applicability of corrective justice, which complements whatever other claims migrants or refugees can make.

7.1. Realizing Obligations under Corrective Justice: Sidestepping the Issues of Causation

The argumentative position of climate migrants is weakened by the issues of causation. Due to the issues of causation, they cannot provide sufficient evidence that, and to what extent, they were really induced to migrate because of anthropogenic climate change. Thus, it stands to reason that short of being a "mere" conceptual problem, the issues of causation are morally and politically challenging issues. But challenging for whom? It appears to be intuitively unacceptable that polluting countries, by pointing simply to the absence of a clear evidentiary link between their particular actions (i.e. emissions) and other people's displacement, can eschew their moral responsibility under corrective justice. But this is precisely what countries can do under current international law: there is a normative gap which allows countries to eschew their responsibility for the fate of climate migrants as long as a clear evidentiary link cannot be established.[1]

In the scarce normative debate on climate migration, the issues of causation equally tend to work to the disadvantage of the climate migrant, shielding emitters from responsibility under corrective justice. The rationale is that if it is not possible to establish developed countries' greenhouse gases as the cause for particular environmental harms, it is difficult to apply a corrective justice frame and thus to hold developed countries responsible for their actions.[2] Often, it is the climate migrant that bears the burden of proof. It is he who has to show —by providing evidence on the causal relationship between the environmental deterioration that impelled him to leave and the putative duty bearer's (i.e. the state's) emissions—that he is entitled to some form of (corrective justice-based) assistance that goes beyond the "mere" human rights-based assistance owed to people in need.

---

[1] See Zetter (2010, p. 139).
[2] See Wyman (2013, pp. 193-95), McAdam (2012, p. 93). And as Posner and Weisbach contend, the issues of causation, though "not fatal to corrective justice claims" (Posner and Weisbach, 2010, p. 110), weaken such claims significantly.

Opposing such an assessment, I will now argue that even when an evidentiary link cannot be fully established for the *specific* case (epistemic issue of causation) — and even when the exact causal contribution relative to other economic or demographic migratory triggers cannot be determined (issue of multi-causality) — the putative climate migrants can make a substantive claim to some form of assistance against putative duty-bearers under corrective justice. The argumentation unfolded in the following sections will support the view that climate migrants need not shy away from the burden of proving other parties' responsibility. They can carry that burden and make — even in the face of the issues of causation — a powerful case for compensation.

7.1.1. The Vantage Point: General Causation as Opposed to Specific Causation

This is the first argumentative step for arguing that — despite the impossibility of demonstrating clear evidence of the specific causal link — climate migrants *do* hold a powerful compensatory claim against putatively responsible parties. The solution to be proposed here draws on the distinction between "general causation" and "specific causation". In a nutshell, the core rationale behind the conception of "general causation" as I will here understand it by drawing on Huggel *et al.* (2016) is that the basic causal mechanics leading to climate migration *are* well understood. We may lack understanding regarding the specific migration-triggering impacts of climate change (that is, specific causation), but there is presumably sufficient understanding of the general processes at work. There is some general understanding regarding what countries and regions contribute most to climate change and what countries and regions (rather than specific individuals) will be affected most by the (migration-triggering) effects of climate change.

> We draw here an analogy to US environmental litigation, where typically two types of causation are relevant: "general causation" refers to the question of whether a substance is capable of causing a particular damage, injury or condition, whereas "specific causation" refers to a particular substance causing a specific individual's injury.[3]

Likewise, we do not know whether and in what ways a specific set of emissions added to a specific migration-triggering climatic event, but it is sufficiently clear that in the sense of general causation that very set of emissions was capable of contributing to that overall process of climate change which in turn is capable of leading to particular migration-

---

[3] Huggel *et al.* (2016, p. 903).

triggering climatic events. The general mechanics that lead from emissions to migration are generally understood.

However, there is in current normative debates on climate migration a dominant focus on what I refer to here as specific causation, and I will argue that this focus is not justified. Instead, an understanding of general causation, centring on the well-understood mechanisms underlying climate change, is the more promising epistemic vantage point.[4] While neglected in current debates on climate migration, the focus on general causation will allow for much clearer reasoning on what we owe to (potential) climate migrants and their affected home regions or home countries. (It will turn out presently that the shift away from specific causation will, at least to some extent, go hand in hand with a shift away from the affected individual and toward the affected region or country.) By taking a conception of general causation as the vantage point for subsequent discussions, it is possible to show that the issues of causation, contrary to what many authors suggest, do not present the high hurdles to the climate migrants' ability to make a forceful special claim against potential host countries.[5]

Let me now try to be a bit more concrete on the nature of the difference between general and specific causation, and on the merits of focusing on general causation. Now, there can be no doubt that from the perspective of (heavily) emitting states, before they can actually accept their responsibility for the (migration-triggering) harms of distant people, it is of course highly relevant to have clarified who caused or contributed to what extent to the negative impacts that impelled those affected distant people to leave their home countries. However, and this is the decisive point that comes with the here proposed shift toward *general causation*, there is no need to establish *one* party as causally responsible for *one* specific migration-triggering event or process.

---

[4] Note that this focus on general causation is relevantly different from a "probabilistic" or "more likely than not" standard that Hodgkinson et al. (2012) adopt in their attempt to make sense of climate migration.

[5] Huggel et al. (2016, p. 903). With a view to the victims of climate harms generally, they argue that the "hurdles are considerable, and they may range from aspects of justiciability, to the proof required for causation, to the applicability of the no-harm rule established in international law or of the application of extraterritoriality in human rights law" (ibid.). Moreover, they observe that, as regards some of the poorest developing countries most in need of assistance, it will be particularly difficult to attribute negative climatic impacts to anthropogenic climate change. For people there, it will thus be particularly difficult to bring forward "any claim for liability and related compensation in international climate policy or at courts" (ibid., p. 907).

To make clearer what is meant by *general* causation in the concrete case of climate migrants, one might construct the following example: it may be unclear whether—in the sense of specific causation—a particular crop failure in a particular country at a particular time can be credited to anthropogenic climate change so that it remains unclear for that specific instance whether and to what extent the (looming) migration of the people affected by that particular crop failure is (also) induced by climate change. What one can observe though is a more general causal link that already provides for some considerable normative thrust: there is evidence of the region being generally exposed to certain climatic changes, there is available some general understanding of the sensitivity of certain crops to climatic changes, there is available some general understanding of the vulnerability of human populations to the economic losses that result from such crop failures, and there is some general understanding of the ways in which such losses, interacting with various other social and political determinants, induce people to leave their homes at some stage.[6] This reasoning would apply to both sudden-onset events and for slow-onset processes. In either case, the here proposed "understanding of general causation [...] can rely on multiple lines of evidence collected from observations, modeling or physical understanding, but not all are necessarily required and nor do they all have to concern the exact impact and location in question".[7]

As already indicated, the move from specific to general causation arguably seems to direct the present discussion's attention toward a higher, collective level and away from a concern with the individual. As Huggel *et al.* observe: "From an attribution point of view, governments are in a better position to claim compensation than individuals, because damages due to climate change can be aggregated over time and space over their territory and/or economic interests."[8] Countries as politically organized can argue more credibly that understanding of causation in the specific case is negligible in light of the fact that the state's territory is large, that it exists through time, and that there can hardly be any serious doubt in the sense of general causation that this territory and its people—"over time and space"—has been affected by the detrimental effects of anthropogenic climate change. However, this focus on states rather than individuals does not imply that only states (for their affected nationals) can make a claim to assistance based on

---

[6] Huggel *et al.* (2016, p. 904).
[7] Ibid.
[8] Ibid., p. 903.

corrective justice against other states. Rather, it could mean in practice that a forced climate migrant from Bangladesh does not have to point to evidence that clarifies the specific way in which she was turned into a climate migrant or the specific way in which she, in so far as she is still a *potential climate migrant*, faces the risk of being displaced at some point. Instead, if an understanding of general causation were made the relevant standard for assessing our obligations toward climate migrants, the affected person would only have to point to the fact that she is from a certain region in Bangladesh—i.e. a region of which one can reasonably assume that it is generally affected by climate change and climate migration—and her claim would assume additional, corrective-justice based normative vigour. In order to determine whether a given country or region should indeed be regarded as affected by climate change and subject to climate-related migratory pressures, one could draw on politically agreed thresholds (which themselves would be informed by data and indicators form the climate sciences), so that the validity of the affected person's claim would then depend on whether or not that threshold has been crossed in her particular locality, country, or region.[9]

One could of course object here that the proposed focus on general causation is politically unfeasible, because the switch from the focus on specific causation to a focus on general causation—which for its operationalization requires governments and politicians to actually agree on that switch—could result in much higher costs for those governments: once (potential) climate migrants could draw on some notion of general causation in order to establish other countries' (corrective justice-based) responsibility, those other countries would presumably see an explosion in costly demands made on them. Now, it may indeed be so that such increased costs could deter putatively responsible states from accepting general causation as the relevant vantage point for assessing whether they are responsible toward a certain group of putative climate migrants or not. But this leaves untouched the decisive moral point: namely that the just proposed focus on general causation provides the aggrieved parties, i.e. the climate migrants, with a substantive argumentative lever for supporting their claim of other parties' responsibility under corrective justice. Whether the putatively responsible parties will accept that lever or not is a secondary concern here. What matters is that, to the extent that the

---

[9] See Maguire (2017, p. 126), who proposes this move in the more specific context of (compensatory) adaptation funding, which I will come to in the next sub-chapter (7.2).

rationale behind the proposed shift toward general causation is morally appealing, putatively responsible countries would come under some moral pressure when they refuse to accept that rationale. That they would indeed come under pressure is not unlikely. For after all, general causation is solidly grounded in our basic understanding of climate change and of the ways in which such climatic changes induce migration. The big losers from climate change can capitalize on that evidence. Even though (potential) climate migrants will invariably fail to establish a clear evidentiary link between their specific instance of migration and the emissions of mainly others, this does not mean that there is no way for them to make their case by drawing on those others' causal role.

Putatively responsible parties cannot hide behind affected parties' inability to establish specific causation. In so far as affected parties often have the lever of general causation on their side, they might ultimately challenge any attempt to allocate to the affected parties alone the burden of proving the assumed causal link. And indeed, one might ask provocatively, why not reverse the burden of proof? Why not have the main contributors prove that they are not responsible in the sense of specific causation when (and in so far as) it is, after all, already clear that they are responsible in the sense of general causation? Such reasoning may be provocative and it will hardly result in practically useful (or feasible) policy advice. But it is morally instructive. It suggests that putatively responsible parties will be forced to leave their moral comfort zones. They are not shed against claims of corrective justice. This claim will be substantiated in the subsequent section.

7.1.2. *The Relevant Wrong: Acquiescence as Disrespectful Conduct against Others' Rights*

For corrective justice to be ultimately applicable there must be observable in reality some actions or some conduct that is relevantly contrary to rights. There must be a detectable invasion by one party of some other party's rights; there must be a conduct that cannot count as minimally *respectful* of others' rights. If any such conduct is observable, some kind of compensation appears to be due. However, simply to hold that any kind or form of emitting GHG *by itself* constitutes such a disrespectful rights invasion that requires compensation would appear to be little helpful and, presumably, morally inappropriate.[10] It is

---

[10] One reason why it may be considered morally inappropriate lies in the constraints that individuals and countries encounter when aiming to stop emitting

clearly the case that emissions combine for the *harmful process* of climate relatedness in which otherwise distant people are connected through emissions and their detrimental effects on some people. But arguably, it would be doubtful to reason back from the recognition of this harmful process and simply refer to any single emitted unit of GHG that added to that overall process as harmful and as contrary to other people's rights.

So what then is it which in the harmful context of climate relatedness constitutes and can be identified as a *disrespectful* conduct that counts as contrary to rights and on which, therefore, claimants could ground their claims to compensation of some kind? Before I set out to identify such conduct, it bears noting how Iris Marion Young contends that in contexts of complex causal relatedness one should not waste one's time trying to make claims to compensation:

> Because the particular causal relationship of the actions of particular individuals or organizations to structural outcomes is often not possible to trace, there is no point in seeking to exact compensation or redress from only and all those who have contributed to the outcome, and in proportion to their contribution.[11]

This view of Young amounts to a capitulation in the face of the issues of causation, and in my own dealing with the challenge of blurry causality I will clearly deviate from this apologetic view of Young. Contrary to her optimistic approach, I think that simply to assume a forward-looking normative perspective that completely replaces any liability model (like corrective justice) is not a convincing solution here. Rather, her approach is an invitation for emitters to keep emitting in a business-as-usual manner. It relies on the optimistic assessment alone that the parties contributing to the injustice will accept what she thinks is their responsibility to counter that very injustice and that they should thusly do away with the structural injustice in a concerted effort in which every participant provides for his or her part of shared responsibility. This approach with its total abdication from the language of corrective justice and compensation shall be rejected here.

---

abruptly and in the real risks to energy security, to economic, and to social stability that states take when withdrawing from fossil fuels too abruptly. These are important considerations that a hasty condemnation of any kind of GHG emissions would appear to leave unconsidered. It is clearly the case that emissions must be reduced to zero as quickly as possible. But to hold that any single act of emitting *as such* constitutes a blameworthy action would appear to be an inappropriate moral assessment of the reality of climate-related harm.

[11] Young (2005, p. 722).

Young deprives herself of a powerful argumentative tool which leaves her approach unnecessarily toothless.

So what then is the conduct which despite the issues of causation is observable in reality and which would count as relevantly contrary to other people's equal rights? Arguably, what is against other people's rights is that emitters *acquiesce* in the possibility of contributing to a process that is generally understood to infringe those other people's rights. Recall how it was argued in the second chapter that a disrespectful conduct can also include someone's imposition of a relevantly severe risk to another person, and for such disrespectful conduct to count again as "*respectful* of the victim's rights"[12] – compensation is required. Now, when emitters add to the risk of harming others through their emissions – and drawing on general causation they have reason to assume that this conduct of emitting indeed adds to the risk of harm – and when those emitters do little or nothing to counter that overall structural injustice or to reduce others' vulnerability to that risk, then it seems they portray a behaviour that will not count as sufficiently respectful of other people's equal rights. Even though in the climate case the harm cannot be easily stopped, the general point remains intact, namely that not to do anything about the continuing harm is disrespectful of the harmed people's rights. And this makes this conduct contrary to rights, which then in principle can ground claims to compensation.

It may be unclear whether the putatively responsible party's actions caused the harm and it may be so that the emissions *as such* hardly qualify as disrespectful of others people's rights. Nonetheless, emitting parties act against those other people's rights when, knowing of the general possibility that their emissions contribute to a process of structural injustice which invades other people's basic right not to be harmed, they simply do nothing to change that structure and keep imposing risks on distant others. At minimum, they can be expected to either work in some way against that unjust structure (mitigation) or to aim at reducing potentially affected people's vulnerability to that unjust structure (adaptation). If they don't, they acquiesce in it, and this acquiescence provides the affected parties with a clearly identifiable conduct that is not sufficiently respectful of their equal rights. On this disrespectful conduct they can then ground their claims to compensation.

As Coleman points out, such compensation then does not address concrete wrongful harmings or rights invasions, but is due simply in

---

[12] Coleman (1992, p. 284).

order for A's "conduct to count as *respectful*"[13] of other people's rights. Importantly, the charge that compensation is in principle due is not weakened by the recognition that the people who claim such compensation are often only potential and putative victims. What matters is that the acquiescent conduct portrayed by emitters fails to show sufficient respect for their rights. In the context of the causal blurriness of climate change, the identification of this conduct that grounds their claims to compensation appears to provide (even potential and putative) climate migrants with a credible and resilient (argumentative) lever which would allow them to dodge the issues of causation. Even when still in situ and even when a certain risk has not yet materialized, they raise claims against (distant) others who failed to show sufficient respect for their equal rights by acquiescing in the contribution to a process which imposes on them a risk (against which they may remain unprotected). How such claims to compensation could be framed in more concrete terms remains to be discussed in the next section.

After this series of two argumentative steps in which (1) I argued for the new focus on general causation and in which (2) I identified *acquiescence* as the relevantly disrespectful conduct in the context of climate migration, the ground for applying corrective justice is almost prepared. But one important question remained open so far: if the conduct on which claims to compensation can be based is *acquiescence*, then what is the *alternative* course of action which would count as "non-acquiescence", which would not (at least not to the same degree) count as disrespectful of others' rights, and which would thusly not call for the application of corrective justice? If the problematic conduct does not lie in the single and clearly identifiable action of emitting a certain unit of greenhouse gases, it remains essentially vague what emitters have to do (in more positive terms) in order for their conduct not to count as disrespectful of others' rights. One purpose of the following section is to respond to this problem of vagueness and to point out what the morally required alternative conduct could amount to. This alternative conduct will be sketched along general lines and it will be argued that to pursue it amounts to a *general preventive responsibility*. The case for this responsibility is by and large a climate-ethical one, which I will build by pulling together the elements developed in the previous sections: namely the focus on general causation and the identification of *acquiescence* as the relevantly disrespectful conduct contrary to other people's rights. In bringing these strands together, I

---

[13] Coleman (1992, p. 284).

will further corroborate my overarching contention that climate migrants *can* draw on corrective justice when claiming assistance.

7.1.3. The Moral Minimum: The General (Preventive) Responsibility to Reduce Others' Vulnerability

The climate context as analysed in previous sections is marked by uncertainty and vagueness. It is not only impossible to pinpoint specific causal relationships between action and harmful outcome. It is also rather unclear what is to be done in concrete terms, and by whom, in order to reduce effectively and in the fastest possible way the level of harm produced in that context of climate relatedness. Of course, one could argue for a simplistic "just stop all emissions right now", but in so far as the present analysis works with realistic assumptions about the constraints emitters face, what can be realistically demanded from them tends to be more ambiguous. To begin with, these ambiguities call for a helpful conceptual distinction—namely that between "responsibility" and "duty". The shift from specific duty to general responsibility as I shall propose it now fits together in a rather natural way with the above focus on general causation as opposed to specific causation and with the emphasis on structural injustice (and disrespectful acquiescence) as opposed to specific harms. This is how Young sets forth the conceptual difference between "duty" and "responsibility":

> When we have a duty, moral rules specify what it is we are supposed to do [...]. Responsibility, however, while no less obligatory, is more open as to what actions it calls for. One has the responsibility to do whatever it takes to bring about specific ends or purposes. Taking responsibility also involves exercising more discretion than enacting a duty does. It is up to the agents who have a responsibility to decide what to do to discharge it within the limits of other moral considerations.[14]

Conceptually, speaking of a "responsibility" as opposed to a "duty" better handles the problems of ambiguity discussed above: that it is not clear what emitters can be expected to do in light of the constraints they face, that it is not clear how to address the structural injustice most effectively given the reality of scarce (financial, political, motivational) means, and—with a view to the basic problem of the issues of causation and attribution—that it remains unclear how the specific actions of the emitters bear on the rights of distant others so that it would be problematic to simply hold the emitters liable for compensation on the mere grounds of their emitting. These considerations render the language of responsibility more suitable: emitters then have a general responsibility

---

[14] Young (2011, p. 143).

not to acquiesce in the (potentially) harmful reality of structural injustice.

In the same way that the above understanding of general causation is tied rather to territorially and temporally extensive units like countries — because extending across time and space they are generally and more easily understood as contributing to and/or as being affected by climate change — the here introduced conception of *general responsibility* is also most plausibly applied at the collective level of countries or regional blocs (like the EU). Indeed, and in so far as in accordance with the "responsibility thesis" in tort law the "scope of one's responsibility is [...] coextensive with the causal upshots of one's volitional conduct",[15] it appears to make sense to apply the conception of general responsibility primarily to larger units like countries, and not to individuals. At that aggregate level, the "causal upshots" of people's emissions in a given country or group of countries combine for a meaningful degree of causal responsibility that can be established so that generally responsible (state) parties can be identified more readily. At that collective level, inaction and business as usual can, depending on the country's or the bloc's size, make the generally understood, harm-producing structural (climate) injustice even worse. And at that politically coordinated level, claims can be addressed to governments (or other bodies) in charge, and effective counter-measures in the form of effective energy or climate policy can get off the ground and make a difference.

What correlates with the basic right not to be harmed is then not a distinctly negative duty or responsibility (I) to *refrain* from some action or course of action. Rather, with the general responsibility being a more open demand not to acquiesce in something that is generally understood to contribute to structural injustice and thus to individual harm, this demand urges the (politically organized collective of) emitters to show a credible commitment in the realms of climate policy, energy policy, technology policy, or in terms of assistance offered to potentially or actually affected parties in adapting to the effects of climate change. In this way, the *general responsibility* stretches across the multi-layered structure of duties correlative of a basic right not to be harmed. In so far as "[t]aking responsibility also involves exercising more discretion than enacting a [specific] duty does",[16] the here proposed conception of a general responsibility has the welcome effect that philosophers can sidestep the near-impossible task of spelling out in

---

[15] See Coleman (1992, p. 273).
[16] Young (2011, p. 143).

detail what it is that the relevant actors in the process of climate change have to do and what they have to forbear from doing. This should of course not bar the philosopher from concretizing how the general responsibility can be cashed out in the more specific case of climate migration (which I will do in the next sub-chapter).

After all then, one may summarize the general responsibility identified and systematized in this section as a hard-to-contest *moral minimum*: what emitters have toward (potential) climate migrants in light of the "mere" possibility that their emissions could add to the structural injustice (and in this way potentially have a detrimental bearing on the realization of other people's basic rights not to be harmed) is a general preventive responsibility to reduce the potential victims' vulnerability to that structural injustice. The mere possibility of contributing to a harmful process urges the emitter to do something to counter the generally understood harmfulness of that process. Note how this finally brings my argumentation full circle: for when emitters fail to live up to (even) this general responsibility, putative victims would find the identifiable wrongful conduct of *acquiescence* which is contrary to their (and other people's) rights and which they can draw on for grounding their claims to compensation under corrective justice. In this way, they retain the possibility of corrective justice even in the face of the discussed issues of causation. They build their case on a particular attitude that accompanies those emissions, namely on a form of acquiescence that is disrespectful of their rights.

Given the issues of causation, given the at best marginal role of individual emissions, and given the constraints the individual emitters would face when urged to stop emitting abruptly and completely, it would appear to miss the mark to simply postulate a specific duty to refrain from all emissions right now. However, and this is a quintessential finding of this section: the facts that people and countries are related through their emissions, that the terms of this relatedness are unfair, and that specific emissions are generally understood to contribute to this overall injustice together combine for a solid enough reason to assume that each and every emitter—most plausibly understood as emitters on the collective level of countries—have a responsibility *to do something* about that injustice, or as Young puts it, they have at least the responsibility "to work to remedy these injustices".[17] They have the general responsibility to do something to achieve that aim, and if they fail to do even that, it is sound to hold them liable for their

---

[17] Young (2005, p. 709).

complacent and irresponsible conduct, which is unquestionably contrary to and disrespectful of other people's equal rights.

7.2. The Responsibility to Forestall (Climate) Migration

In the previous section, it was argued that *acquiescence* in the possibility of (further) contributing to structural injustice constitutes a conduct that is in a relevant sense contrary to others' equal rights, and it was then pointed out that the alternative conduct to such wrongful acquiescence is (in the sense of a moral minimum) to live up to one's *general (preventive) responsibility* to reduce the vulnerability of those (distant) people who are potentially affected even by one's non-wrongful contribution to the structural injustice of climate change. This combined insight is the starting point of this sub-chapter. Here I will claim that—as a concretization of the general responsibility—what is first and foremost owed to (potential) climate migrants (under corrective justice) is a form of ex-ante compensation that aims at reducing their vulnerability to that very process which could induce them to migrate in the first place. As a key observation, such ex-ante compensatory measures make sense in the climate case because climate change is generally characterized by a *time lag* between the potentially harmful actions (i.e. emissions) and the actually detrimental effects.[18] Note further that ex-ante compensation should be seen to fall into the realm of adaptation: I take it to encompass those measures which are adopted before (i.e. ex-ante) rather than after (i.e. ex-post) people are negatively affected by certain detrimental climate-related impacts. A central contention will be that such ex-ante compensatory measures include, among other things, the establishment of an insurance scheme which insures the (potentially) affected against climate risks, which renders them less vulnerable to those risks, and which thusly reduces their proneness to (forced) migration.

7.2.1. The Responsibility to Compensate Ex-Ante

The focus now turns from a concern with the general process of climate change back to the more particular case of climate migrants. As regards this particular case, the general preventive responsibility not to acquiesce and thus not to disrespect others' rights shall be framed as a form of *ex-ante* compensation to those people mainly exposed to climate risks. All of the people exposed to such risks are in principle, i.e.

---

[18] See Baatz (2017, pp. 146, 212). As Baatz points out, it is this time lag that allows for an intervention before the negative effect has fully unfolded.

conceptually speaking, *potential climate migrants* who remain in situ, and compensating them ex-ante will turn out to be the most effective way of assisting them.

Ex-ante compensation is understood here as concretization of the *general responsibility* to counter the harmfulness of the process of climate change. As already suggested, it is meaningful to speak of "ex-ante" (or preventive) compensation in the climate case because there is generally a time lag between the potentially harmful (and, as the case may be, disrespectful and wrongful) action of emitting greenhouse gases on the one hand and the actually harmful events, processes, or changed climatic conditions on the other hand. This time lag allows responsible parties to live up to their responsibility in advance of the potential materialization of the risk to which their conduct contributed. It creates a window of opportunity for anticipating, easing, or even avoiding the possible harm: ideally, the changed climatic conditions are prevented from unfolding their negative impacts on the realization of human rights in the first place. While ex-post compensation replaces a good from which some right-bearer was deprived, ex-ante compensatory measures aim at avoiding the necessity to replace that good in the first place, which expresses a central idea of adaptation. With regard to what may seem to be the alternative to a thusly understood ex-ante compensation—namely to wait for the risks to materialize and then provide for *ex-post* compensation—to provide for ex-ante compensation is clearly the morally preferable course of action.[19] After all, the here proposed conception of ex-ante compensation is concerned with the possibility of future harm and with countering that future harm. So rather than with actual and past harms (as *ex-post* compensation would be), it is concerned with *risks* of harm and with reducing people's vulnerability to that risk.

Although motivated by actions against rights, ex-ante compensation is grounded not in actual losses or setbacks to the substances of a right which a right-bearer suffers due to another person's conduct. Rather, and in line with the focus on risks, claims for ex-ante compensation are grounded in another kind of conduct against rights: namely in the disrespect a party portrays in view of the rights of others. It was pointed out before that acquiescence in the structural injustice of climate change is in this sense disrespectful of others' equal rights. It is

---

[19] For this rather commonsensical assessment that preventive (i.e. ex-ante) compensation—which aims at forestalling and thusly at avoiding the need for replacing a good—is preferable to more conventional forms of ex-post compensation—which aims at replacing a good—see Goodin (1989). Note though that Goodin does not use the terminology of ex-ante and ex-post compensation.

contrary to rights not primarily in the substantive sense that the substance of another person's right is tangibly impaired, but rather in the more abstract, temporally prior, and rather procedural sense that insufficient respect is shown for another person's *prima facie* right not to be exposed to a risk of suffering a severe setback to the substances of his rights in the first place. This is the diagnosis on which the claim for ex-ante compensation grounds. To the extent that putative duty-bearers provide for such ex-ante compensation, their prior conduct of emitting will count as at least less disrespectful of the (potential and putative) victim's rights.[20] In this way putative duty-bearers anticipate and reduce the thrust of whatever *later* claims to ex-post compensation putative climate migrants might (or might not) raise against them.

The most important reason why ex-ante compensation should be preferred to ex-post compensation draws on one of this work's central conceptual findings: namely on the (temporal) *fluidity* of the categories of climate migrants. Recall that it was proposed earlier to understand the phenomenon of climate migration as a process that invariably begins in the affected people's home region (i.e. in situ), where they can already be conceptualized as *potential climate migrants*. To the extent that environmental conditions are gradually deteriorating in that region, those *potential climate migrants* are increasingly likely to become *actual* climate migrants, and whether they then figure as "climate-induced migrants" or "forced climate migrants" (or "climate displaced persons") depends essentially on the *timing factor* of their movement (e.g. whether they wait for the water to be up to their knees or not). Against the backdrop of this fluidity, it was argued that in a relevant sense the *climate-induced migrant* presents at least *potentially* the same moral claim against us as the *forced climate migrant* (because the arbitrariness of the timing factor should hardly make much of a moral difference to us) and by the same rationale one can now contend that the *potential climate migrant* while still in situ also and already holds potentially the same claim to assistance against us. And when it is generally understood that a group of persons will be better off the earlier we assist them and when this group holds a moral claim to assistance against us, then this understanding should normally urge us to act earlier (i.e. ex-ante) rather than later (i.e. ex-post), *all other things being equal*.

---

[20] I build here on the reasoning provided by Coleman, who points out that (and in what sense) compensation may sometimes be "required in order for an injurer's conduct to count as respectful of the victim's right" (Coleman, 1992, p. 92).

In addition to this general reason in favour of preferring ex-ante (preventive) measures over ex-post compensation, the fact that anticipatory adaptive action tends to be more cost-effective also strongly supports the case for discharging our obligations toward climate migrants earlier rather than later. As one research group found, every dollar that is invested in preventive adaptive measures on average reduces climate-related damages by about four dollars.[21] Such findings lead Baatz (2017) to the conclusion that "allocating a given amount of resources to (ex ante) adaptation rather than rectification might be more efficient in safeguarding possibilities to exercise human rights."[22]

The best of the three options of a) staying in situ, b) migrating, and c) being forced to move is presumably a). In this vein, I argued above for a concentration on the category of *potential climate migrants* and for in situ assistance, for only the effectuation of the right to stay truly reaches to the needs of *all* affected people. It appears to be the most efficient course of action to help people realize option a) and thus their right to stay; and the proposed focus on ex-ante (compensatory) measures resonates with that very insight.[23] Or should those countries rather wait until c) is the only remaining option and then compensate the affected people *ex-post*? This course of action would only extend the failure to respect minimally those affected people's rights. Rather than wait for the affected people to end up as *climate-induced migrants* or as *forced climate migrants*, putatively responsible countries should try to *forestall such claims to ex-post compensation* by discharging their (corrective justice-based) responsibility by means of providing for ex-ante compensatory measures.

But if countries fail to compensate ex-ante, i.e. before emigration is triggered, there appears to evolve a potential reason why climate migrants (who may at some stage figure as *actual* climate migrants) could make a *special claim* to being admitted to a potential host country, and the existence of this special claim could seriously compromise the reasonableness of exerting the *right to exclude* as I formulated it in the third chapter. In such cases where the affected migrants come from a region that is *generally understood* to be affected by migration-triggering climate harms and if ex-ante compensatory measures were not provided in their case, it will appear that the exclusion of such (generally understood) climate migrants would be a most doubtful thing to do.

---

[21] Hof *et al.* (2009, p. 837).
[22] Baatz (2017, p. 213).
[23] For the observation that assistance in situ tends to be the most efficient (i.e. cheapest) and (therefore) the morally required one, see also McAnney (2012, pp. 1187–90).

Once those affected people reach such countries as actual climate migrants, the latter could point to an (uncompensated for) *detrimental relationship* between them and the host country, and this *detrimental relationship* would ground a *special claim* (under corrective justice). By the above rationale underlying the formulated right to exclude, it is this uncompensated and generally understood *detrimental relationship* (or, as the case may be, any more *specific* harmful relationship) that would render the *exertion* of the right to exclude a rather doubtful thing to do. Excluding them would extend a prior injustice done to them. This is indeed an important insight gained from the analysis so far.

7.2.2. Instituting an Insurance System: Countering Climate Migration

A concrete way of discharging one's responsibility to compensate *ex-ante* consists in working toward the institution of an *insurance* system that aims at rendering *potential climate migrants* less vulnerable to those climate risks which, where they to materialize, could induce them to migrate. Insurance is an effective adaptive measure that strengthens resilience in affected areas: "By providing financial security against droughts, floods, tropical cyclones and other forms of weather extremes, insurance instruments present an opportunity for developing countries in their concurrent efforts to reduce poverty and adapt to climate change."[24] In many developing countries affected by climate change, where insurance premiums are not affordable to the poorest members of society, there is a shortfall in such climate catastrophe insurance so that one way of assisting the affected countries could consist in setting up insurance schemes and in subsidizing premiums.[25] In this section, I will present reasons why insurance against climate risks can be an effective measure for addressing the challenges of climate migration raised in this work. Needless to say, to discharge one's responsibility by providing (or rather: by subsidizing) insurance is only one form of assistance that should be combined with other risk-reduction measures. But as I will point out, insurance is particularly suitable as a response to many of the specific challenges of climate migration as raised before.

The idea is that putatively responsible countries put in place and pay premiums (in some sense proportionate to their share of global emissions)[26] into an insurance system as a means of discharging their

---

[24] Linnerooth-Bayer *et al.* (2009, p. 381).
[25] See Lyster (2015, p. 426).
[26] Note that the proposal that premiums be determined depending on a given country's (present and past) GHG emissions resonates with the common-sensical suggestion that, whatever funds there are to address the challenges of

responsibility to compensate ex-ante. The insurance, led by expert knowledge, then uses these funds in order to reduce *potential climate migrants'* vulnerability, which it does both (1) by helping to adapt before a risk materializes and (2) by assisting after a risk materialized. In this way, although the funds were provided by the putatively responsible party as a means to discharge the responsibility in an ex-ante manner, the way in which these funds are made available and allocated may, from the aggrieved party's perspective, take shape as ex-post compensation. As Baatz observes, an "interesting feature of insurance is that it combines ex post and ex ante elements: a risk reduction/pooling scheme is set up before the impact occurs but remedy is provided after its occurrence".[27] Aware of their *entitlement* to such remedies in the event of harm, the aggrieved party will be, and will have the *perception* of being, ultimately less vulnerable to climate risks. This reduction in the perception of risk is essential.[28] For as pointed out before, it is such subjective risk perceptions that play a decisive role in the individual decision process behind (eventual) migration. This insight is further evidence for the adequacy of the insurance approach for tackling climate migration in an effective and sustainable way.

The adaptive approach of insuring the people affected by detrimental climate impacts is already mentioned in the 1992 United Nations Framework Convention on Climate Change. There, it says that:

> [T]he Parties shall give full consideration to what actions are necessary under the Convention, including actions related to funding, insurance and the transfer of technology, to meet the specific needs and concerns

---

climate migration, contribution to such funds should be determined on the basis of states' common but differentiated responsibility. This "common but differentiated responsibility" principle can be found at the heart of the treaty establishing the United Nations Framework Convention 1992, in which the parties to the convention acknowledge "that the global nature of climate change calls for the widest possible cooperation by all countries and their participation in an effective and appropriate international response, in accordance with their common but differentiated responsibilities and respective capabilities and their social and economic conditions".

[27] Baatz (2017, p. 221).
[28] For the central role that perceptions of risks play in individual migratory processes, see Hugo (2012). He points out that "it will not just be the real or experienced hazards or depleted resource condition that leads to increasing emigration, but also the changes in *perceptions* towards environmental risk or resource degradation" (Hugo, 2012, p. 15). Recall that my above categorization of climate migrants takes account of this important role of risk perceptions in individual migration processes.

of developing country Parties arising from the adverse effects of climate change and/or the impact of the implementation of response measures [...].[29]

Indeed, a thusly proposed insurance scheme has come to be reflected in international institutional practice in so far as there is by now a series of diverse financial mechanisms and funds that aim in one way or another at helping vulnerable countries or regions adapt to the detrimental impacts of climate change or at compensating for the losses that already occurred. For example, under the umbrella of the United Nations Framework Convention on Climate Change (UNFCCC) and thusly as part of the UN climate regime, there has been established the "Warsaw International Mechanism for Loss and Damage associated with Climate Change Impacts" which aims at addressing "loss and damage associated with impacts of climate change, including extreme events and slow onset events, in developing countries that are particularly vulnerable to the adverse effects of climate change".[30] And there already exist within the complex architecture of the UNFCCC such funds as the Least Developed Country Fund (LDCF), the Adaptation Fund (AF), and other (smaller) funds.[31] In so far as those distinct funding streams already pursue, in different ways, the purpose of reducing the risk exposition of different (vulnerable) addressees, they already at least partly capture the idea of an insurance scheme. However, none of such finance mechanisms explicitly refers to its risk reduction efforts as a means of *insuring* the affected against loss and damage.

What I propose is that in line with such achievements and tendencies in practice, an insurance scheme should figure (more explicitly and more visibly) as a central pillar in a broader system of adaptive measures under the UN's climate regime.[32] As already indicated above, the UN's climate regime has already and tacitly started to assume its role in governing climate displacement.[33] For pragmatic

---

[29] UNFCCC (1992, Article 4.8).
[30] UNFCCC (2014b).
[31] See UNFCCC (2014a).
[32] This view is also held by Maguire (2017).
[33] Maguire predicts that, rather than any other body, it is the UN and its climate regime that will play the decisive role in assisting people displaced by climate change (Maguire, 2017, pp. 121–25). In so far as the UN climate regime is also concerned with the current and historical emissions of its parties, the UNFCCC appears to be a suitable forum in which parties can be urged morally—for example by linking up presumed failures in the realm of mitigation with demands in the realm of loss and damage—to help vulnerable people affected by the migration-triggering effects of climate change.

reasons and because the UN appears to be in a good position for generating substantial financial funds, it thusly makes sense to call for the UNFCCC to become the central actor in governing climate migration. (This institutional separation would allow the UNHCR to concentrate its efforts on the protection of political refugees, at least to the extent that this conceptual distinction is possible in the concrete case.) An effective insurance system — intended as one way for states to discharge their responsibility to assist (III-1 *and* III-2) — will help struggling states in fulfilling its duty to protect (II).

Now, conceptually, and with a view to this chapter's ongoing and persistent quarrel with the issues of causation, the idea of instituting an insurance system as a concrete way of dealing with the demands of climate migrants appears to be a promising strategy. Indeed, my previous argument for sidestepping the issues of causation now culminates in this proposal of insuring climate migrants. An insurance system that is placed between the putatively responsible emitters and the potential (but hardly identifiable) victims of such emissions is well suited to deal with the uncertain causal relationships much discussed in this work. As Peter Penz — a defender of the approach to insure climate migrants — puts it: compensation (where it is due) to presumed climate migrants "on a no-fault basis means that disentangling complicated webs of causal links is avoided".[34] And indeed, as a related point stressed by Penz, aggrieved parties should have their damages covered even in those cases where "detailed demonstration of causal responsibility"[35] is not possible. In this sense, the call for an insurance system further substantiates the case for *sidestepping* the issues of causation.

But here one could object that such an insurance scheme could leave contributing parties paying for damages in excess of what they can be held causally responsible for. This worry is a consequence of the issue of multi-causality: beside the climatic factor, there will generally be several other factors that combine for the particular severity of a given loss — like social or political context factors that distant emitters are hardly responsible for. For example, beside the climatic factors, the specific severity of a flood could be attributable "to a combination of poor urban planning, deforestation, lack of floodwater management systems and failure of previous master plans on flood mitigation".[36] Parties adding to the insurance scheme will hardly be willing to pay for

---

[34] Penz (2010, p. 167).
[35] *Ibid.*, pp. 172–73.
[36] Lyster (2015, p. 427).

those "parts" of the losses that they can hardly be made responsible for. As it seems, the issue of multi-causality could complicate the establishment of the proposed insurance scheme. So how to deal with the tendency that most losses tend to be the result of the interplay of several economic, social, political, *and* environmental factors?

Arguably, the described problem could be solved by making the insurance an index-based one: the aggrieved party gets payouts from the insurance only when a previously defined weather-related threshold is passed: "The essential feature [...] is that the insurance contract responds to an objective parameter (e.g. measurement of rainfall or temperature) at a defined weather station during an agreed time period."[37] Thresholds should be defined in such a way that they reflect only those weather events and climatic processes that are presumably not the normal in the region, but a likely consequence of climate change. So only when the severity and frequency of, for example, droughts or hurricanes overshoots a threshold that likely would not have been trespassed without climate change, then — and only then — payouts would be triggered.[38] As Baatz observes in a related context, payouts "are not based on actual losses but on over- or undershooting a threshold".[39] This index-based character responds to the issue of causation. Moreover, in so far as the insurance would often cover only a portion of the incurred losses, it would appear to do justice to the conceptual difficulty of multi-causality.[40] And after all, as the putatively responsible parties that set up the insurance know in advance that they will normally not pay for all of the damages, the here proposed insurance scheme would appear to become more politically feasible.

However, such thresholds appear to be of at least limited use when it comes to sudden-onset disasters. In such cases, it might remain essentially unclear whether a specific materialized risk also would have materialized in the absence of anthropogenic climate change (although, in principle, the threshold could here be defined in terms of the *frequency* of certain extreme sudden-onset events). As regards such remaining cases, I propose with Penz that "when in doubt, we should err on the side of inclusion rather than exclusion of disastrous events and compensation for them."[41] Finally, another possible problem with an index-based insurance system is that the considered parameters and thresholds — while ideally informed by scientific understanding alone —

---

[37] International Fund for Agricultural Development (2011, p. 15).
[38] See Baatz (2017, pp. 222–24).
[39] *Ibid.*, p. 21.
[40] For this assessment, see also Baatz (2017, p. 21).
[41] Penz (2010, p. 168).

could become the object of political bargaining processes. This should be avoided by the putatively responsible parties involved in setting up the insurance scheme.

Through the here proposed insurance system, the vulnerability of the potentially affected parties is effectively reduced: they may still suffer the immediate damage, but they will have it (partly) compensated for, which means an objective reduction of vulnerability. Their resilience is enhanced and their adaptive capacity increased, which ideally decreases the affected people's risk of being *forced* to move.[42] As already indicated, however, the insurance should not use its funds exclusively to assist people who have already suffered from the materialization of a risk. Rather, it should also take preventive action that aims at reducing the likelihood or severity of possible damages from occurring in the first place, for example by investing in the resilience and adaptive capacities of (potentially) affected communities.[43] Such anticipatory measures are required both to counter slow-onset environmental processes and to increase resilience against sudden-onset disasters. With a view to the specific concern with climate migration, such measures aim at countering the above described tendency that affected people gradually move along the continuum from the status-quo category of *potential climate migrant* toward the ultimate stage of *forced climate migrant*.

Note further that the call for an insurance scheme with its focus on *potential climate migrants* also resonates with the earlier insight that only the fulfilment of the *right to stay* truly reaches to the needs of the poorest people in the source countries.[44] It covers also those poorest and weakest members of the affected countries who would be unable to migrate even if legal channels were provided. It is in this sense that the fulfilment of a right to stay makes sure that *all* affected people are effectively covered. An ethical approach that discusses climate migration entirely by asking what we owe to climate migrants in terms of immigration-related responsibilities would invariably remain blind to the claims and the fate of the larger group of people exposed to climate-induced *triggers* of migration. In contrast, an insurance approach with

---

[42] The insurance scheme (as one central adaptive measure to combat climate migration) thusly responds to the view of authors like Warner *et al.* (2013), who conclude that the central aim of any adaptive measure to cope with climate migration must be to increase resilience at the household level and by that means to render the movement of affected people a *choice* (ibid., p. 16).
[43] See McMichael *et al.* (2010).
[44] For this emphasis on the right to stay, see also Hugo (2010, p. 26) and Barnett and Webber (2010, p. 40).

its focus on adaptation measures in situ covers those people. It aims at protecting all affected people—the actually and the potentially affected—against the standard climate threats. It aims at effectuating the *right to stay*.[45]

Needless to say, even if an effective insurance scheme were in place, it would not suffice to fully protect the affected people against the climate threats they face. In many cases, it will hardly be possible to prevent (forced) migration. Sometimes, it will be just too late or simply not possible to prevent people from migrating. It is in such cases that the insurance's funds should also be used for migration-related forms of adaptation. Insurance payouts would then not only be used for financing and managing the affected persons' migration, but would also go to communities that consent to receive the migrants and pay the costs of accommodation. As Penz puts it: "If cross-border migration turns out to be the best option, then funds may need to go to receiving states, both as an incentive to accept migrants and as compensation for expenses incurred in accommodating and integrating them."[46] In many instances, however, internal rather than international migration could turn out to be the more suitable option so that the facilitation of such internal movement should also be considered when deciding on the usage of funds in the specific case.[47]

At any rate, when the primary strategy of protecting people against risks in situ fails, any form of governance must be in place to supplant that primary strategy. Whether this task is eventually fulfilled or coordinated by the same insurance scheme as I just proposed, or whether it is the UNHCR that steps in then,[48] or whether existing

---

[45] For such a rather comprehensive insurance system to be effective, it would, presumably, have to be harmonized with other (UN) spheres of governance alongside the sphere of the environment, for example the sphere of development (e.g. the UN Development Programme (UNDP)). A discussion of the merits of such an integrated scheme is provided by McAdam (2012, pp. 212-19).

[46] Penz (2010, pp. 170-71).

[47] In fact, it is often argued that the "Guiding Principles on Internal Displacement", an existing (but non-binding) soft-law instrument in international law, is relevant and readily applicable to the context of climate migration, especially in so far as the Guiding Principles could cover that significant share of climate migrants that will not cross an international border. For this assessment, see Lyster (2015, p. 435), Kolmannskog (2012, pp. 39-43) and McAdam (2012, pp. 250-52).

[48] Needless to say, for the UNHCR to cover forced climate migrants on top of the refugees and internally displaced persons it already takes care of would require that it profoundly reorganizes itself, as Lyster (2015, p. 436) submits. Such an extension of the UNCHR's mandate would come with the risk that people in

norms in international law can be applied or extended are important questions that legal theorists, advocacy groups, and policy makers will keep discussing.[49] From a moral standpoint, there can be no doubt that some legal instruments must be in place in order to assist affected people when the primary strategy of insuring them against the migration-triggering climate risk failed.

Arguably, this legal instrument should not consist in a *climate refugee treaty* though. It should not consist in the establishment of a distinct legal category of "climate refugees" which would then compete with other categories like Convention refugees.[50] The above proposal for an insurance scheme and its integration into a broader institutional framework would, if instituted effectively and supported determinedly by putatively responsible parties, avoid the unworthy competition of different groups of needy forced migrants for scarce entrance places. The recognition that the climatic factor is inextricably linked up with other economic, social, and political factors calls for a more holistic, integrative, and coordinated approach to the management of global migratory flows. An insurance scheme resonates with this insight: the usage of its funds can be attuned to measures taken in other spheres of governance, and it can largely dispense with the idea of enshrining people affected by climate migration into fixed *legal* categories to which certain material entitlements are tied. The rather comprehensive category that the insurance scheme is predominantly concerned with is the *potential climate migrant* and more generally still: vulnerable people in regions identified as subject to migration-triggering climate threats.

---

host societies become less willing to accept the higher burdens (i.e. the higher demand for immigration) that such an extended mandate would tend to imply. But this reduced acceptance or even anti-immigration sentiment is clearly a more general problem associated with higher numbers of worldwide forced migrants internationally on the move; it would equally apply if an institution other than the UNHCR managed the additional flows of (forced) climate migrants. As already suggested repeatedly, this insight provides for a pragmatic reason in favour of preventive and in situ measures as such measures would come with lower political and financial costs.

[49] Such discussions can be found, for example, in McAdam (2012), Epiney (2011), Biermann and Boas (2010), and Cournil (2011).

[50] For a balanced account of the ongoing debate on the usefulness of a treaty on "climate refugees", see McAdam (2012, pp. 186–210). McAdam, while favourable of the idea of environmental migration governance, remains sceptical of a treaty solution. Other authors like Piguet *et al.* (2011), Cournil (2011), or Stavropoulou (2008) similarly highlight the specific difficulties related with such a global treaty approach. For a more favourable stance, see Biermann and Boas (2010), Hodgkinson *et al.* (2012), Williams (2008), and Docherty and Giannini (2009).

Although some effort was made in the previous analysis to point out that corrective justice supports this case for building up an insurance scheme, it is important to recall that the normative principle of corrective justice is—even though a very powerful one—only one normative source alongside others. To the extent that the here proposed insurance scheme presents an effective instrument for securing the basic rights of people threatened by the migration-triggering effects of climate change, and to the extent that it can be assumed that the threatened people's home states cannot provide for the protection (II) of those rights, such an insurance scheme appears to be warranted just as well by a concern (III-1) with the basic rights of the vulnerable people in the affected regions.

Indeed, this case for assistance based on basic rights and the mere ability[51] to provide for that assistance becomes even stronger if one points to the generally understood circumstance that those countries which could help, i.e. which have the financial means to set up such an insurance scheme and to pay premiums generously, often have those means—*among other things*—because of their very greenhouse gas emissions today and in the past.[52] It is clear who benefitted most from those emissions that cause climate change, and it is clear who has the means to assist the aggrieved parties. The general point I am making here is that these additional considerations could be adduced to complement and make stronger the case for assistance based on corrective justice. Together, these complementary normative strands converge toward one powerful argument for assisting (potential) climate migrants.

Finally, one should once more caution against the view that an insurance scheme is by itself sufficient to solve the problems related with the phenomenon of climate migration. It is only one pillar in a broader adaptive scheme, and that broader international scheme should itself be backed up (and informed) by other concrete measures and forms of institutional governance at the *regional* level. As Jane McAdam observes, there is some "considerable activity at the regional level to address climate change-related movement".[53] Such regional

---

[51] In the climate ethical literature this rationale figures as the *Ability to Pay Principle*.

[52] This rationale is referred to as the Beneficiary Pays Principle (BPP), and it appears to converge here with the Ability to Pay Principle. For a more general account of how several independent climate ethical considerations tend to converge toward the conclusion that some countries are particularly responsible for countering climate change, see Shue (2015).

[53] McAdam (2012, p. 233).

(and indeed local) responses could turn out to be indispensable tools at least until a more internationally integrated scheme is negotiated, and even then such regional instruments could remain and would have the important advantage of being more sensitive to specific local forms of climate migration. They would be more sensitive to the specific demands of climate migrants in a given region. On the regional level, better empirical understanding to address and anticipate the needs of climate migrants could be generated.

But what if putatively responsible countries disrespect their responsibility to provide for ex-ante compensation and fail to forestall the affected people's eventual migration? Arguably, they would then be obliged to provide the affected parties with ex-post compensation. But as I will point out in the next sub-chapter, measures provided as a means of ex-post compensation will almost invariably remain inappropriate. Such measures cannot make good for the specific losses that were suffered. If this claim can be established, it would further strengthen the case for fulfilling the general responsibility to provide for ex-ante compensation.

7.3. The Inappropriateness of Ex-Post Compensation and the Pitfalls of Arguing for Collective Resettlement

> The principle is to return to a state of affairs that is as close as possible to the status quo ante, assuming that itself was not unjust. A similar logic applies to a chain of events that creates refugees. Ideally, the responsible state should try to engineer conditions that would enable those affected to return to their previous lives rather than move them to entirely new surroundings. Sometimes repair is impossible, in which case granting the refugees the right to remain permanently in [a certain host country] may be an acceptable, albeit second-best, alternative.[54]

With a view to Miller's reasoning in the above quote I will contend that the states responsible for providing ex-post compensation to climate migrants, i.e. after failing "to engineer conditions that would enable those affected to return to their previous lives", will often fail again in their attempt to provide "appropriate compensation" through admission. A substantiation of this claim and a discussion of its morally-normative implications are the themes of this sub-chapter.

To begin with, it is important to establish an understanding of what *appropriate* compensation would consist in. Arguably, and as suggested by Miller in the above quote, "appropriate" compensation means that form of compensation which leaves the aggrieved party as close as

---

[54] Miller (2016, p. 115).

possible to her *status quo ante*. As regards cases of (climate) displacement, in which the suffered loss is a loss of home, the *status quo ante* would presumably refer to the aggrieved party's former home and more generally speaking to the former context in which she lived and in which she conceived of her life plans before her displacement. Now, a relevant and meaningful criterion for determining how close or distant a given climate migrant is to this *status quo ante* is arguably the degree of her dispositional freedom in her new place of living, i.e. the degree to which she eventually finds herself again in a position in which she can make free choices in accordance with her (previously conceived) life plans and in which she can lead *her* life. So if a person is displaced from her home country due to climatic factors (as we assume) then the thing that would presumably have to be provided in order to compensate her is first and foremost to allow her to move to another country or to resettle her to that other country and then, in that country, to provide her with a new home. But could such ex-post compensation really qualify as an appropriate form of compensation?

There is a fundamental reason for negating this question. Arguably, immigration into another country or resettlement to another country will almost invariably fail to replace what the affected person actually lost. She lost more than her home in the material sense. Instead, she lost her *context of choice*, and it appears to be almost impossible to engineer those conditions that will replace that *context of choice*. As Avner de Shalit (2011) observes, the place and context that climate migrants are forced to leave is constitutive of their identity because of the way it bonds them to their "values, history, personal and collective memory, language, natural surroundings, to things we are familiar with and at ease with".[55] In this way, it provides them with "a sense of belonging to something greater than ourselves individually. It offers a sense of home".[56] Based on this view Avner de Shalit argues that "compensation misses the essence of the loss associated with environmental displacement."[57] A sense of home, the claim goes on, is *incommensurable*, so that there is after all no way in which one can appropriately compensate the victims of forced climate migration.

The Dutch family in Greece, and more clearly still the Tuvaluan family in Finland, will lack the particular *context of choice* they had in their home place. They will not find replaced that context on which they drew for formulating their life plans and on which much of the

---

[55] De Shalit (2011, p. 318).
[56] *Ibid.*
[57] *Ibid.*

realization of these life plans depended, i.e. on the natural surroundings, on the values, on societal institutions, and finally on the people with whom they shared this context and which together shaped it. This suggests that it is in many cases difficult and often impossible to fully compensate climate migrants through granting them (and effectuating) individual rights to immigrate, to be resettled, and after all to a secure and free life in another country. But is there an alternative *feasible* way of compensating them? What—if not individual immigration or resettlement—could be a more suitable way for bringing victims back to their *status quo ante*?

As some authors propose with a view to people who have been displaced in consequence of sea-level rise and the associated *disappearance* of entire states, the appropriate form of compensation is *collective resettlement* rather than individual resettlement.[58] One possible rationale behind such a proposal is this: if individually resettled climate refugees can hardly be fully compensated in other cultural and political contexts because they there lack a particular *context of choice* to which they attach value and which cannot be replaced, then one solution to this difficulty could indeed be to take the whole issue of compensation to a collective level. For if the problem with fully compensating (forced) climate migrants (or "climate refugees", as the here discussed authors simply call them) lies in the fact that their *status quo ante* is so closely tied to their particular political communities' *contexts of choice* and if this context cannot be replaced through individual immigration or resettlement, then—these authors suggest—one should instead try to resettle the political community as a whole.[59] In other words, one would by that view have to provide (forced) climate migrants with new territory on which they can, together with the other climate migrants from their community, re-establish their form of life.[60] A decisive implication of this argument for countries that failed to fulfil their preventive responsibilities is that ex-post compensation is expensive. To relocate a community is costly and can imply the cession of territory.

Dietrich and Wündisch (2014) argue for precisely this (costly) scheme of collective relocation and they base it on a "theory of pure territorial compensation".[61] They stress "that a community which has lost its territory due to anthropogenic climate change is under no obligation to adapt to new surroundings. Since its members have a claim to full compensation, they cannot be expected to significantly

---

[58] See, for example, Risse (2009) and Kolers (2012).
[59] See e.g. Mathias Risse (2009) and Dietrich Wündisch (2014).
[60] See Dietrich and Wündisch (2014, pp. 13-14).
[61] *Ibid.*, p. 12.

change their traditional cultural practices."[62] The individual and the collective level converge here toward a normative case for collective resettlement: the individual has a right to nothing less than full compensation and thus to a life in his own community, and the political community *as a collective* has a right to exercise its self-determination in a "culturally meaningful way".[63] As the two authors point out: "A political self-determination right protects a group from becoming subject to rules which are hostile to its particular culture. Hence, a compensatory territory must allow the resettled population to exercise cultural activities to which they attach great value."[64] In so far as Dietrich and Wündisch limit their case to the inhabitants of low-lying island states at risk of being flooded (a limitation that, in the following, remains to be criticized in itself), the overall number of political communities to be relocated collectively would, at least on their account, appear to remain rather small.

There are several problems with Dietrich and Wündisch's argumentation. As already indicated, one problem is that they make their claim only with regard to the displacement of the inhabitants of low-lying islands, which could indeed turn out to be a rather specific and relatively rare instance of climate displacement. This limitation obviously compromises the applicability of their normative findings to the wider group of (forced) climate migrants and people displaced by climate change generally. Can this limited focus on such a specific and small sub-group be justified? There is reason to be sceptical. In order to justify this limited focus one would have to make the assumption that there is an "important difference between individual displacement due to climate catastrophes and the disappearance of entire states". This is precisely the assumption Dietrich and Wündisch take for granted. They subscribe to Kolers' (2012) claim that the harm suffered by a refugee who, *together with the collective of his political community*, has lost his entire state through sea-level rise is greater than the loss suffered by someone who is *individually* displaced from a country which after his displacement continues to exist. For it is in Kolers' view only in the case of the person losing her entire state together with its political community and the self-determination it once exercised, that her own

---

[62] Ibid., p. 14.
[63] Ibid.
[64] Ibid.

political identity, her culture, her language—or in short: her former *context of choice*—are lost.[65]

This view is hardly convincing though. A climate migrant who is (ultimately) forced to leave his country in consequence of desertification, extreme weather, or continuous flooding will take little comfort in the fact that other more resilient or geographically better placed members of his former political community still manage to live in that country or that its territory continues to exist. Of course, it may make some psychological difference for the individual if her former political community and its territory literally ceases to exist through sea-level rise, which for example would make eventual return altogether impossible. But the crucial point is that in either case the individual will —invariably—suffer a loss of place, a loss of her *context of choice*. Dietrich and Wündisch are thus too quick in assuming a categorical difference between these two instances of forced climate migration. They offer no answer to the question of how to compensate other (forced) climate migrants who—and this they do not see—suffer a loss that is equally severe and equally difficult to compensate. Those forced climate migrants (or climate displaced persons) have just as much a right to *full and in-kind* compensation. Territorial compensation is on Dietrich and Wündisch's account reserved to the small group of people fleeing small island states, which is indeed a comfortable view to take. If, however, one assumes that all forced climate migrants who cannot remain in situ have such a right to full in-kind compensation, and the reasons provided in this section support this view, then the irresponsible states which failed to live up to their responsibility to mitigate and compensate ex-ante would get under pressure morally and territorially. The group of people who could make a forceful claim to territorial compensation could be much larger than the relatively small number of people fleeing small island states.

However, as Campbell presumes, it is rather unlikely that even a collectively relocated group could truly "sustain its 'way' in a foreign land"[66] and it is, moreover, completely unrealistic that responsible states would ever cease correspondingly large parts of their own territory in order to make possible such relocation on the required scale.[67] But the theoretically more important problem is this: the majority of

---

[65] See Dietrich and Wündisch (2014, p. 11), and for the position on which they themselves draw here and which they support (at least in this relevant respect), see Kolers (2012, p. 334). For a similar view, see Moore (2015, p. 210).

[66] Campbell (2010, p. 67).

[67] As Zetter observes, large-scale resettlement so far remains a relatively rare "feature of the protection discourse" (Zetter, 2012, p. 136).

climate migrants will flee from states whose territory and institutions *continue to remain in place*, for it is often "only" and primarily the more vulnerable groups of society that will be forced to move. This observation implies that collective relocation will be technically *impossible*. For if people do not flee as a collective then those who flee cannot be compensated and resettled as a collective. Their state, their institutions, and their *context of choice* cannot be *re*-placed on a new territory because it is still, at least partly, *in* place at its original place. The upshot is that this (much larger) group of forced climate migrants can hardly be fully compensated in the way Dietrich and Wündisch propose with respect to the rather small group of people fleeing small low-lying island states. So after all, the recognition that it will be impossible for a large number of climate migrants to be granted what they have a right to — namely full in-kind compensation — reinforces the case for preventing climate change and forestalling climate migration. For as de Shalit notes, states should not "assume that they can let global warming happen and then [...] compensate for it".[68] From a risk-ethical perspective, the fact that the risk — where it materializes — cannot be compensated for (appropriately) is a most decisive recognition that urges us to prevent, or at least reduce others' vulnerability to, the risk in the first place.

7.4. Outlook: The Consequences of Irresponsibility

The insights gained in the previous sub-chapters call for a few rather general concluding remarks: in a time of increasing global disorder and crisis, the number of people in need of assistance could well increase. This is a bleak outlook indeed. For our means are limited, not only for assisting them in situ, but especially for assisting them by admitting them.[69] With such troubled times looming, responsible countries had

---

[68] Avner de Shalit (2011, p. 325).
[69] Walzer (1983, p. 51). One could ask here whether putatively responsible countries should grant (forced) climate migrants preferential admission over other non-climate migrants in times when there are too many migrants (or refugees) asking for entrance. Indeed, it could be argued that when a selection among forced migrants cannot be prevented because the number of those needy people asking for entrance far exceeds the number of available places, responsible countries should — *all else being equal* — indeed prefer forced climate migrants over other forced migrants toward whom they do not bear an *additional* corrective-justice based responsibility (if, despite the *epistemic issue of causation* and the *issue of multi-causality*, this distinction can be made in the concrete case). This would be an uncomfortable conclusion. The outlook of a future world in which those refugees whom we "caused" ourselves are the luckiest ones is unsettling.

better remain in a strong argumentative position; they are well-advised to fulfil their preventive responsibility and their responsibility to compensate ex-ante. After all, this is the only way for them to anticipate and hopefully prevent (at least partly) that actual climate migrants will at some point make a hardly deniable special claim to admission against them. If countries fail to fulfil those responsibilities, they will face difficulties in morally defending certain more restrictive immigration-related measures they otherwise deem necessary. If a putatively responsible country failed to show adequate *respect* for those people's rights and if the country failed to *compensate* (ex-ante) for that disrespect in order to render it less disrespectful, then not to admit climate migrants will at some point hardly seem like a morally acceptable option.

So what options remain? At some point, it seems, responsible countries would have to allow for more immigration; and this requires that they provide for those conditions in society that allow the (additional) migrants to become members who participate as political and economic equals. As Fareed Zakaria (2016) suggests:

> [T]he crucial element in the mix is politics: countries where mainstream politicians have failed to heed or address citizens' concerns have seen rising populism driven by political entrepreneurs fanning fear and latent prejudice. Those countries that have managed immigration and integration better, in contrast, with leadership that is engaged, confident, and practical, have not seen a rise in populist anger.[70]

These final remarks set a new tone. They seem to eventually deviate from the realistic "political approach" taken so far. But the distinctly realistic rationale behind this move is this: should the more pessimistic estimates on the quantitative dimensions of climate migration become reality, then determined policies to foster among citizens a higher degree of open-mindedness toward the issue of immigration would indeed figure as the last hope to prevent a moral catastrophe of large numbers of (climate) migrants remaining without a place of shelter. This final appeal amounts to the recognition that once large-scale climate migration is a reality in a warming world, the readiness of people in potential destination countries to accept (and indeed endorse) a more liberal migration regime will remain the only morally acceptable way forward.[71] This is no solution, but the necessary "concession"

---

[70] Zakaria (2016, p. 14).
[71] So departing from the assessment of Zakaria, it appears that how particular societies respond to the intake of large numbers of further migrants and refugees will in large part depend on the measures such societies take to alleviate the effects of immigration. They could, for example, try to cushion the

(if open-mindedness can ever be that) *after* countries failed to solve the issue by *anticipating* it. The alternative to this ultimate concession would be to close the borders, to securitize climate migrants, and to leave them to their own fate, if they allow this to be done to them.

---

negative economic effects that immigration tends to have for some existing groups in society, like the working poor. Such measures could help render attitudes toward immigration more positive.

Eight
Conclusion

The results I have come to in the course of this work can be structured by dividing them up into three themes: (i) community and exclusion, (ii) humanity and refugeehood, and (iii) climate change and responsibility. After summing up the insights gained with a view to each of these themes and after pointing out how these insights should inform normative reasoning on immigration in the 21st century, I will in this conclusion criticize very shortly the peculiar tendency in current normative debates on climate migration of focusing on the (overly) specific case of sea-level rise (iv). Moreover, I will reflect once more on the merits of the "political approach" that I chose in this work (v). At the end of this conclusion, I will indicate open questions for future research (vi).

(i) Community and Exclusion

The right to exclude was shown to be an important authority that states should have in order to preserve a secure *context of choice* which individuals have an essential interest in and which secures nothing short of the functioning of democratic institutions. This argumentation for the importance of culture and community was derived from my account of basic rights and thusly differs radically from the way in which culture figures in some communitarian accounts or from the way in which Eastern European countries (or Western populist politicians) misuse culture in order to defend exclusive national identities and toxically restrictive immigration policies. The upshot of my discussion on the importance of community is captured in Alan Gewirth's verdict that "rights require community for their effectuation, and community requires rights as the basis of its justified operations and enactments."[1] Democratic states *need* the right to exclude in order to protect and maintain those very conditions that allow them to fulfil their primary task of effectuating the (basic) rights of their citizens; and the citizens of democratic societies have that right to exclude because it was them

---

[1] Gewirth (1996, p. 97).

who have been contributing to the establishment, functioning, and proceeds of those institutions and they therefore appear to have a right to control access to those institutions.

The strategic rationale of this first bloc on the effects of immigration, on the importance of community and finally on the right to exclude was to provide for that scenery and general awareness against the backdrop of which the later discussion on climate change would unfold its full normative thrust. Indeed, the recognition that a secure *context of choice* and social trust among citizens are valuable goods that can erode through certain forms of immigration proved to be indispensable for cautioning emphatically against a looming context in which the legitimacy of *exerting* the right to exclude could become seriously challenged.

At any rate, my approach stands in contrast with the cosmopolitan case for global free movement, which I rejected early on. The calls for open borders in the ethics of immigration remain surprisingly insensitive to the reality that the functioning of democratic institutions can be compromised by certain levels of immigration, not least because some members of society are—like it or not—reluctant to engage in deliberative democracy with people with whom they have little in common. The justified worry is that if cosmopolitan elites push successfully for an open borders regime and if they declare any more restrictive immigration policy as incompatible with human rights or even as implicitly racist,[2] then those parts of society who identify less with such cosmopolitan ideals will feel overlooked. Right-wing populists will receive them with open arms. Indeed, as even liberal commentators like Fareed Zakaria admit, "Western societies will have to focus directly on the dangers of too rapid cultural change. That might involve some limits on the rate of immigration and on the kinds of immigrants who are permitted to enter."[3] Many cosmopolitan authors in the ethics of immigration, on the other hand, appear to be unwilling to make such concessions or to admit just how seriously such concessions would compromise their case for open borders. They will remain unworldly as long as they keep operating at their present distance from the real (and not necessarily racist) worries of people who remain hesitant as to the possibility of maintaining liberal institutions in the face of large numbers of immigrants, especially when they come from culturally distant places.

---

[2] See Collier (2013, p. 22).
[3] Zakaria (2016, p. 15).

(ii) Humanity and Refugeehood

As regards the special case of refugees, I defended the view that we have very strong duties to assist refugees. With their basic rights at stake, they make a very strong *claim* to admission against the potential host country. Diverging from definitions of refugeehood in international law, it was assumed that a sufficient criterion for granting a right to asylum is *that* a person is deprived of the substances of basic rights and that it is irrelevant *why* she is deprived of them. Moreover, the moral claim of the refugee was shown to become almost undeniable when the host country contributed through its prior actions to his refugeehood in the first place and when it is thus responsible under corrective justice. These insights on the morally required dealing with refugees would prove to have some bearing on the question of what is owed to (forced) climate migrants.

Just as importantly, however, my elaborations on the special case of refugees also had the purpose of making a rather independent contribution to current ethical debates on refugee policy. In my approach to the question of when "our" duties toward refugees reach their limits, I criticized some authors' unwillingness to deal with this issue head-on. The insight that the affected people are deprived of basic rights cannot mean that we have an almost endless duty to keep taking in refugees, i.e. almost to the point of such assistance becoming self-defeating. By drawing on the empirical understanding of the dynamics behind immigration and its tendency to strain fragile cooperation games within society, I contended that a risk-ethical approach is best equipped to deal with this delicate question of where to draw limits. Only a risk-ethical perspective is sufficiently sensitive to the reality that the causal relationship between the number of admitted refugees and the eventually resulting strains on the functioning of societal institutions is generally a dynamic and indeed an uncertain one. If it is uncertain where the limits of absorptive capacities lie and if it is thus uncertain at what point efforts in refugee policy could turn out to be self-defeating, then the issue must certainly be couched as a risk-ethical one and it is surprising how little this has been attempted in current normative debates. Any reliance on verdicts like "ought implies can" or "ultra posse nemo obligatur" then turns out to be oversimplistic in the context of immigration and refugee policy. They cannot inform policy making as long as there is a lack of understanding concerning the extent to which the helper actually *can* help.

This discussion on limits was meant to take account of the real-world developments that were raised already in the introduction to this work: namely that immigration and refugee policy is a truly divisive issue which—as the ongoing European refugee crisis makes painfully clear—can have (or already has) disruptive societal and political effects. In this heated context, a concern with the limits of our responsibilities could well prove to be a vital one. The recognition that certain regions of the world experience unsustainable population growth only renders this debate on limits more urgent.

Hope lies in the insight that assistance by admission is only one way of discharging one's responsibility and that, on most reasonable assumptions, well-conceived efforts to assist in situ promise to be the more efficient and morally preferable form of assistance. This assessment would become particularly relevant for the question of how to deal with climate migration.

(iii) Climate Change and Responsibility

My concern with the phenomenon of climate migration would put to the test the previously established normative framework. After pointing out the climatic triggers of migration, my concern with climate induced migration quickly turned toward the specific conceptual problems that characterize climate migrants. A first challenge, I argued, consists in coming to terms with (1) the particular difficulty of categorizing climate migrants and of tying specific moral entitlements to those respective categories. The upshot of my conceptual analysis was that it often depends only on the arbitrariness of the timing factor (behind a particular migrant's decision to move) whether this affected person is, when assessed from a synchronic perspective, grasped as belonging to the first category of voluntary *climate-induced migrants* or to the second category of *forced climate migrants* (or the third category of *climate displaced persons*). The recognition of essentially fluid categories was shown to make nonsense of the conventional practice of granting special rights to assistance or admission only to people figuring as *forced* migrants. Especially in so far as slow-onset environmental degradation is concerned, I concluded that even those who in the particular moment at which they reach a potential host country would still figure as voluntary *climate-induced migrants* because they emigrated considerably *before* their region became inhabitable can make a most powerful claim to admission: their basic rights are insecure in the medium or long-term so that they cannot be seen to enjoy dispositional freedom in their home country. I proposed that a partial response to this problem of categorizing climate migrants lies in putting the focus on a category that is far less slippery, namely the *potential climate migrant*.

8. Conclusion

The other major set of conceptual problems were grouped under (2) the issues of causation. Here, the most far-reaching problem was the following: if it is often not possible to tell with certainty whether and to what extent a particular migrant is indeed a climate migrant because the causal relationships between emissions of GHG and migratory processes remain essentially blurred as regards the specific case (*epistemic issue of causation and attribution*), then it seems there will be no way of making practical sense of the intuition that climate migrants are special and somehow deserving in practice of "our" special moral attention. This difficulty of speaking meaningfully of climate migrants is further complicated by the fact that the climate is in most cases only one causal factor among other environmental, social, economic, or political contributing factors (*analytical issue of multi-causality*).

Concerning the issues of causation, and the *epistemic issue* in particular, a series of two argumentative steps was developed that aimed at providing a strategy for sidestepping the difficulty of establishing a causal link between emissions and migration for the specific case. First, I argued for focusing on *general causation* as opposed to specific causation, and then I contended that acquiescence in the generally understood harmfulness of climate relatedness and thus in the possibility of harmful climate migration be regarded as the relevantly *disrespectful* conduct contrary to other people's rights on which the affected can ground a *special claim to assistance or admission under* corrective justice. This decisive finding that—despite the issues of causation—there is in principle a responsibility under corrective justice eventually made further practical sense of the intuition that we have special obligations toward climate migrants. More still, I found that for a country to exclude (forced) climate migrants after all possibilities to prevent their migration and to compensate them *ex-ante* have been foregone would be a morally most doubtful thing to do. It would mean to extend a series of injustices that were previously done to them.

It turned out that both assessments, namely that (1) the focus should be on *potential climate migrants* in situ and (2) that potential host countries are responsible under corrective justice, powerfully converge toward the verdict that what we have toward (potential) climate migrants is a strong *preventive responsibility*, which is often suitably framed as a responsibility to compensate the affected people *ex-ante*. Such ex-ante compensation was shown to be an option because of the time lag between the (disrespectful risk imposition through) emissions and the materialization of the (possible) harm.

As a promising way of institutionalizing these forms of responsibility especially toward *potential climate migrants* I argued for an insurance scheme that collects premiums from responsible parties and uses these funds first and foremost in order to reduce vulnerability in situ by strengthening resilience of local communities. It is only when in situ assistance is no longer possible in the particular instance that the funds should also be used to facilitate migration (as adaptation) as the second-best alternative. The proposal for an insurance pillar (for example within the UN's climate regime) as one form of assuming one's responsibility was shown to respond to the issues of causation in so far as it makes unnecessary the demonstration of causal responsibility for the specific case. Premiums are collected in accordance with a country's share in global emissions and payouts are given to (putatively) aggrieved parties only in regions which are *generally understood* to be affected by climate harms and only when certain parameters trespass a predefined threshold, for example when the severity and frequency of hurricanes overshoots a threshold which in the considered region likely would not have been trespassed without climate change. In this sense, the call for an insurance system further substantiated the case for sidestepping the issues of causation. Moreover, the insurance approach was also shown to take account of the *issues of causation*: payouts are triggered only when certain thresholds are passed, and responsible countries should not end up paying for (those parts of the) losses that other factors than the climate account for—e.g. economic, social, or political factors.

It was proposed that this insurance scheme be integrated into the existing United Nations' climate regime or the UN's broader institutional framework and to complement or back up this scheme with other adaptive measures and policy development at the regional level. While in situ reduction of risk exposure and, just as importantly, the reduction of the *perception* of risk exposure were argued to be the priority for policy development, it goes without saying that, where such in situ measures fail, other forms of governance must be in place to supplant in situ assistance. There can be no doubt that legal instruments must be in place in order to deal with those people who were not sufficiently insured against climate risks and who were therefore induced to migrate. Their migration must be facilitated as a means of helping them adapt, and it must be prevented that they end up in an unworthy competition over scarce entrance places with other categories of forced migrants, like Convention refugees. However, there remains reason to worry that if new forms of legal protection are in place for the group of climate migrants and if that group gets too large, anti-immigration sentiment could rise in potential host countries and, in response, such countries could shirk their moral responsibilities toward forced migrants at large.

The case for assisting people in situ and preventing their migration is hardly a ground-breaking one, nor is the idea of risk reduction a fundamentally new one. The contribution of my analysis lies rather in the way in which this case for in situ insurance was shown to answer to the discussed conceptual problems and in which it was shown to flow unambiguously from an account of the equal *rights* of the affected. In terms of rights, the primary aim of the insurance scheme is to protect the rights to dispositional freedom of all people unprotected against the standard threat of climate harms. The aim is to build climate-resilient communities in situ, and thus to effectuate for all of them as well as possible the *right to stay*.

After my account of what (preventive) responsibilities "we" have toward climate migrants, the attention turned toward the moral and practical consequences of not living up to that responsibility. It was pointed out that allowing climate migrants to immigrate will often not compensate them appropriately—a circumstance which I argued to be very problematic from a moral, and not least from a risk-ethical, standpoint. When today's emitting countries fail to take responsibility, they will after all suffer a loss in credibility and moral legitimacy, which are two anchors of stability in an increasingly unsteady world. It is not overly pessimistic to assume that such unsteadiness would even increase in a world with millions of further people forcibly on the move due to climate change. With anti-immigration sentiment already on the rise today, the prospect of a further dramatic increase in worldwide migratory flows becomes an even drearier one indeed. Pure self-interest and the moral arguments unfolded in this work converge here toward a powerful imperative to prevent that dreary future.

(iv) The Questionable Focus on Sea-Level Rise in Current Ethical Debates

There is a peculiar shortcoming in the scarce ethical debate on climate migration that clearly deserves some attention at last: why is it that some of the more influential contributions to the ongoing ethical debate on climate migration concentrate on the overly specific issue of people fleeing from sea-level rise?[4] Why do they put the primary focus on this relatively tiny group and leave the larger group aside? Why do they discuss the very specific duties toward this sub-group of forced climate

---

[4] For articles that remain (more or less) restricted to the special case of sea-level rise, see Adelman (2016), Pellegrino (2014), Risse (2009), Kolers (2012), Dietrich and Wündisch (20140, and Nine (2010).

migrants before there is any proper understanding of the kind of obligations we have toward climate migrants in general? This focus on the more specific group of small island "climate refugees" causes the following suspicion: the focus is laid on them because they constitute the only easily discernible instance of forced climate migration. As Margaret Moore admits, "beyond the most obvious cases, such as submerged islands, it's very difficult to distinguish the drivers of economic migration from climate change more generally."[5] So let me put the mentioned suspicion more trenchantly: the reason why some authors have been focusing on the case of people fleeing from small island states could be that this specific form of climate displacement constitutes a temptingly clear-cut instance of climate migration, a rewarding and welcome playing field for philosophers to apply their theories. In light of my quarrel with the issues of causation in previous chapters, it should be clear by now that simply to eschew the difficulty of causation by restricting one's discussion to the only causally clear instance of climate migration is a problematic and potentially self-deceiving way of "sidestepping" the issues of causation. It leaves the fate of the majority of climate migrants unaddressed.

(v) Reflections on the Political Approach

Even though I proposed in this work practical ways for dealing institutionally with the nexus of climate change and migration, there remains of course the possibility—or indeed likelihood—that responsible parties will eventually not act on those proposals. The preventive responsibility could be neglected by too many countries, or their efforts may simply come too late. It is not difficult to imagine that catastrophic global climate change is not prevented *and* that countries do not manage to develop policy and institutional measures to respond to the by then more acute problem of (large-scale) climate migration.

Now, precisely because I took a *political approach* in the previous analysis, and precisely because I was therefore able to work with the expectation that people might at different stages not do what they ideally have to do (they may *not* stop emissions, they may *not* forestall climate migration by assisting in situ *as long as that is an option*, and they may *not* be persuaded to become more tolerant and accept much higher levels of immigration), I am now in a position to take the final step of such realistic reasoning and open up a rather sinister perspective: on the one hand, the real possibility of large numbers of people unprotected by their own incapable states, and on the other hand potential

---

[5] Moore (2015, p. 210).

host countries which will decide not to allow this increasing number of (climate) migrants to immigrate—for reasons that are under normal conditions very persuasive. This pessimistic scenario should not be dismissed as simplistic or alarmist. It raises awareness for the real risk of human tragedy, improbable though optimists may think that tragedy to be. It is by reasoning back from this sinister scenario that the decisive appeal makes its final appearance: namely to prevent such a moral catastrophe in the first place by mitigating climate change and by forestalling climate migration.

(vi) Open Questions and a Final Imperative

The previous analysis highlighted the need for further ethical reasoning in several fields. Various major and minor remaining questions were already pointed to in the course of the argumentation. Here are two very important such questions which went beyond the scope of my own argumentation and which I consider of particular practical urgency:

- When debating the question of where to draw a limit to the number of refugees admitted, I proposed to adduce a risk-ethical perspective in order to take account of the circumstance that it is generally uncertain when exactly the point is reached at which the admittance of further refugees would strain (fragile) societal institutions just too much. To be sure, a risk-ethical perspective already informed my analysis in decisive respects. But it remains to be specified how one could operationalize such a risk-ethical take on immigration and refugee policy in a more fine-grained way. It also remains to be specified what social scientific informational basis a risk perspective should draw on in the context of immigration policy. Progress on these questions is all the more urgent in a time when the number of forced migrants is on the rise.

- Another delicate question left for future research is how security concerns should be integrated in ethical reasoning on immigration and refugee policy. The recent European refugee crisis has shown that the immigration of terrorists is not a made-up concern of anti-immigration politicians. So what role should such security concerns play in debates on the morality of immigration control? Currently, there is a clear mismatch between on the one hand the lack of serious attention paid to security in the ethics of immigration[6]

---

[6] In none of the recent and more comprehensive contributions to the ethics of immigration, e.g. Carens (2013), Fine and Ypi (2016), Miller (2016), or Wellman and Cole (2013), is security of particular concern.

(apart from downplaying it) and on the other hand the real-world debates and policies that result from precisely this concern. It is important that security get a more systematic place in normative reasoning on immigration. For the alternative is to leave the field to others, namely to an often inconsiderate *securitization discourse* in public debates and in the social sciences.

Apart from these two specific questions, I presume that in a more general sense my discussion of the nexus of climate change and migration invites further reasoning. What is required is a more lively debate on that nexus in normative philosophy. Regarding the possibility that the larger part of worldwide migrants and refugees could at some stage in the near future be climate-induced should be reason enough for contributors to the ethics of immigration to turn toward this pressing issue.

The growing body of research on climate migration from the natural, the social, and the legal sciences must be drawn on by philosophers today in order to build the much needed normative understanding for policy makers to work with. Time presses, not only because of the tendency in both the literature and policy circles to securitize the issue of climate migration. The problem is rather that the scenarios conjured up by contributors to the securitization discourse are, after all, far from pure invention. That a world warmed by more than 2, 3, or even 4 degrees Celsius will not be an especially peaceful place to live is beyond doubt: "Once we hit mass famine, mass migrations and widespread war, the game is lost [...]."[7] At that stage, the above mentioned risk of human tragedy might no longer apply to climate migrants alone. Depending on how many people arrive at our shores, we may not be able to keep at bay — by building higher fences — the problems to which we contributed.

---

[7] Dyer (2011, p. 60).

Bibliography

Abizadeh, Arash (2008) Democratic theory and border coercion: No right to unilaterally control your own borders, *Political Theory*, 26 (1), pp. 37–65.
Adelman, Sam (2016) Climate justice, loss and damage and compensation for small island developing states, *Journal on Human Rights and the Environment*, 7 (1), pp. 32–53.
Adler, Matthew D. (2007) Corrective justice and liability for global warming, *University of Pennsylvania Law Review*, 155 (6), pp. 1859–1867.
Alesina, Alberto, and Glaeser, Edward L. (2005) *Fighting Poverty in the US and Europe: A World of Difference*, New York: Oxford University Press.
Alesina, Alberto, and La Ferrara, Eliana (2005) Who trusts others?, *Journal of Public Economics*, pp. 207–234.
Anderson, Benedict (1983) *Imagined Communities: Reflections on the Origins and Spread of Nationalism*, London: Verso.
Anderson, Elizabeth S. (2009) What is the point of equality?, *Ethics*, 109 (2), pp. 287–337.
Angeli, Oliviero (2015) *Cosmopolitanism, Self-Determination and Territory: Justice with Borders*, New York: Palgrave Macmillan.
— (2015) Das Recht auf Einwanderung und das Recht auf Ausschluss, *Zeitschrift für Politische Theorie*, 2 (2), pp. 171–184.
Arendt, Hannah (1968) *The Origins of Totalitarianism*, San Diego, CA: Harvest Books.
Baatz, Christian (2016) *Compensating Victims of Climate Change in Developing Countries: Justification and Realization*, Greifswald: unpublished PhD thesis (on file with the author).
Bader, Veit (2002) Praktische Philosophie und Zulassung von Flüchtlingen und Migranten, in Märker, Alfredo, and Schlothfeld, Stephan (eds.) *Was schulden wir Flüchtlingen und Migranten?*, pp. 143–167, Wiesbaden: Westdeutscher Verlag.
Banting, Keith, and Kymlicka, Will (2004) Do multiculturalism policies erode the welfare state?, in Van Parijs, Philip (ed.) *Cultural Diversity*

versus Economic Solidarity, pp. 227–284, Brussels: Deboeck Université Press.

Barbieri, Alisson Flávio, and Confalonieri, Ulisses (2011) Climate change, migration and health in Brazil, in Piguet, Etienne, Pécoud, Antoine, and de Guchteneire, Paul (eds.) *Migration and Climate Change*, pp. 49–73, New York: Cambridge University Press.

Barnett, Jon (2011) Human security, in Dryzek, John S., Norgaard, Richard B., and Schlosser, David (eds.) *Oxford Handbook of Climate Change and Society*, pp. 267–277, Oxford: Oxford University Press.

Barnett, Jon, and Webber, Michael (2010) Migration as adaptation: Opportunities and limits, in McAdam, Jane (ed.) *Climate Change and Displacement: Multidisciplinary Perspectives*, pp. 37–55. Portland, OR: Hart Publishing.

Beitz, Charles (1979) *Political Theory and International Relations*, Princeton, NJ: Princeton University Press.

— (2009) *The Idea of Human Rights*, Oxford: Oxford University Press.

Bettini, Giovanni (2008) Climate barbarians at the gate? A critique of apocalyptic narratives on "climate refugees", *Geoforum*, 45, pp. 65–74.

Betts, Alexander (2013) *Survival Migration: Failed Governance and the Crisis of Displacement*, London: Cornell University Press.

Betts, Alexander, and Loescher, Gil (2010) *Refugees in International Relations*, Oxford: Oxford University Press.

Betts, Alexander, and Collier, Paul (2017) *Refuge: Transforming a Broken Refugee System*, London: Penguin.

Biermann, Frank, and Boas, Ingrid (2010) Preparing for a warmer world: Towards a global governance system to protect climate refugees, *Global Environmental Politics*, 10 (1), pp. 60–88.

Black, Richard (2001) Environmental refugees: Myth or reality?, *(UNHCR) New Issues in Refugee Research, Working Paper 34*.

Blake, Michael (2001) Immigration, in Frey, R.G., and Wellman, Christopher H. (eds.) *A Companion to Applied Ethics*, pp. 224–237, London: Blackwell.

Boas, Ingrid (2015) *Climate Migration and Security: Securitisation as a Strategy in Climate Change Politics*, New York: Routledge.

Bodvarsson, Örn B., and Van den Berg, Hendrik (2013) *The Economics of Immigration: Theory and Policy*, 2, New York: Springer.

Boehm, Peter (2012) Global warming: Devastation of an atoll, in Leckie, Scott, Simperingham, Ezekiel, and Bakker, Jordan (eds.) *Climate Change and Displacement Reader*, pp. 408–410, New York: Earthscan.

Borjas, George (1999) *Heaven's Door: Immigration and the American Economy*, Princeton, NJ: Princeton University Press.

Brezger, Jan (2016) So viele wie nötig und möglich! Die Pflicht zur Aufnahme von Flüchtlingen und die Spielräume politischer Machbarkeit, in Grundmann, Thomas, and Stephan, Achim (eds.) *"Welche und wie viele Flüchtlinge sollen wir aufnehmen?" Philosophische Essays*, pp. 57–69, Stuttgart: Reclam.

Brock, Gillian, and Blake, Michael (2015) *Debating Brain Drain: May Governments Restrict Emigration?*, Oxford: Oxford University Press.

Brown, Oli (2008) *Migration and Climate Change*, IOM Migration Research Series, Geneva: International Organization for Migration.

Buchanan, Allen (1999) Recognitional legitimacy and the state system, *Philosophy and Public Affairs*, 28 (1), pp. 46–78.

— (2004) *Justice, Legitimacy, and Self-Determination: Moral Foundations for International Law*, New York: Oxford University Press.

Buhaug, Halvard, Nordkvelle, Jonas, and Bernauer, Thomas (2014) One effect to rule them all? A comment on climate and conflict, *Climate Change*, 127 (3), pp. 391–397.

Cafaro, Philip (2015) *How Many Is Too Many? The Progressive Argument for Reducing Immigration into the United States*, Chicago: University of Chicago Press.

Campbell, John (2010) Climate-induced community relocation in the Pacific, in McAdam, Jane (ed.) *Climate Change and Displacement: Multidisciplinary Perspectives*, pp. 57–79, Oxford: Hart Publishing.

Caney, Simon (2010) Climate change, human rights, and moral thresholds, in Gardiner, Stephen M., Caney, Simon, Jamieson, Dale, and Shue, Henry (eds.) *Climate Ethics: Essential Readings*, pp. 163–177, Oxford: Oxford University Press.

Carballo, Manuel, Smith, Chelsea B., and Pettersson, Karen (2008) Health challenges, *Forced Migration Review*, 31, pp. 32–33.

Carens, Joseph (1988) Immigration and the welfare state, in Gutman, Amy (ed.) *Democracy and the Welfare State*, pp. 207–230, Princeton, NJ: Princeton University Press.

— (1992) Migration and morality: A liberal egalitarian perspective, in Barry, Brian, and Goodin, Robert E. (eds.) *Free Movement: Ethical Issues in the Transnational Migration of People and Money*, pp. 6–22, London: Harvester Wheatsheaf.

— (2013) *The Ethics of Immigration*, Oxford: Oxford University Press.

Cassee, Andreas (2016) *Globale Bewegungsfreiheit. Ein philosophisches Plädoyer für offene Grenzen*, Berlin: Suhrkamp.

Castles, Stephen, de Maas, Hein, and Miller, Mark J. (2015) *The Age of Migration: International Population Movements in the Modern World*, 5, Basingstoke: Palgrave Macmillan.

Clark, William A.V. (2008) Social and political contexts of conflict, *Forced Migration Review*, 31, pp. 22–23.

Coleman, James S. (1988) Social capital in the creation of human capital, *American Journal of Sociology*, 94, pp. 95–120.

Coleman, Jules L. (1992) *Risks and Wrongs*, Cambridge: Cambridge University Press.

Collier, Paul (2013) *Exodus. Immigration and Multiculturalism in the 21st Century*, London: Penguin.

Cournil, Christel (2011) The protection of "environmental refugees" in international law, in Piguet, Étienne, Pécoud, Antoine, and de Guchteneire, Paul (eds.) *Migration and Climate Change*, pp. 259–287, Cambridge: Cambridge University Press.

Cronin, Adian A., Shrestha, Dinesh, and Spiegel, Paul (2008) Water — new challenges, *Forced Migration Review*, 31, pp. 26–27.

De Shalit, Avner (2011) Climate change refugees, compensation and rectification, *The Monist*, 94 (3), pp. 310–328.

Di Fabio, Udo (2005) *Die Kultur der Freiheit*, München: C.H. Beck.

Dietrich, Frank, and Wündisch, Joachim (2015) Territory lost — climate change and the violation of self-determination rights, *Moral Philosophy and Politics*, 2 (1), pp. 83–105.

Docherty, Bonnie, and Giannini, Tyler (2009) Confronting a rising tide: A proposal for a convention on climate change refugees, *Harvard Environmental Law Review*, 33 (2), pp. 349–403.

Donnelly, Jack (2013) *Universal Human Rights: In Theory and Practice*, 3, Ithaca, NY: Cornell University Press.

Doyle, Timothy, and Chaturvedi, Sanjay (2011) Climate refugees and security: Conceptualizations, categories, and contestations, in Dryzek, John S., Norgaard, Richard B., Schlosberg, David (eds.) *The Oxford Handbook of Climate Change and Society*, pp. 278–293, Oxford: Oxford University Press.

Dummett, Michael (2001) *On Immigration and Refugees*, London: Routledge.

Dustman, Christian., Glitz, Albrecht, and Frattini, Tommaso (2008) The labour market impact of immigration, *Oxford Review of Economic Policy*, 24 (3), pp. 477–494.

Dustman, Christian, Frattini, Tommaso, and Preston, Ian P. (2013) The effect of immigration along the distribution of wages, *Review of Economic Studies*, 80 (1), pp. 145–173.

Dyer, Gwynne (2011) *Climate Wars: The Fight for Survival as the World Overheats*, Oxford: Oneworld.

Edwards, Scott (2008) Social breakdown in Darfur, *Forced Migration Review*, 31, pp. 23–24.

Elliott, Loraine (2010) Climate migration and climate migrants: What threat, whose security?, in McAdam, Jane (ed.) *Climate Change and Displacement: Multidisciplinary Perspectives*, pp. 175–190, Oxford: Hart Publishing.

Emanuel, Kerry (2005) Increased destructiveness of tropical cyclones over the past 30 years, *Nature*, 436, pp. 686–688.

Epiney, Astrid (2011) "Environmental refugees": Aspects of international state responsibility, in Piguet, Étienne, Pécoud, Antoine, and de Guchteneire, Paul (eds.) *Migration and Climate Change*, pp. 388–414, Cambridge: Cambridge University Press.

Finkelstein, Claire (2003) Is risk a harm?, *University of Pennsylvania Law Review*, 151 (3), pp. 963–1001.

Foresight (2011) *Migration and Global Environmental Change: Final Project Report*, London: The Government Office for Science. Available at: https://www.gov.uk/government/uploads/attachment_data/file/287717/11-116-migration-and-global-environmental-change.pdf (accessed January 16, 2017).

Fukuyama, Francis (1995) *Trust: The Social Virtues and the Creation of Prosperity*, London: Hamish Hamilton.

Gardiner, Stephen M. (2010) A perfect moral storm: Climate change, intergenerational ethics, and the problem of corruption, in Gardiner, Stephen M., Caney, Simon, Jamieson, Dale, and Shue, Henry (eds.) *Climate Ethics: Essential Readings*, pp. 87–98, Oxford: Oxford University Press.

Gasper, Des (2005) Securing humanity – situating "human security" as concept and discourse, *Journal of Human Development*, 6 (2), pp. 221–245.

Gewirth, Alan (1978) *Reason and Morality*, Chicago: Chicago University Press.

— (1996) *The Community of Rights*, Chicago: Chicago University Press.

Gibney, Matthew J. (2004) *The Ethics and Politics of Asylum: Liberal Democracy and the Resonse to Refugees*, Cambridge: Cambridge University Press.

Gilman, Nils, Randall, Doug, and Schwartz, Peter (2011) Climate change and "security", in Dryzek, John S., Norgaard, Richard B., and Schlosberg, David (eds.) *The Oxford Handbook of Climate Change and Society*, pp. 251–265, Oxford: Oxford University Press.

Goldin, Ian, Cameron, Geoffrey, and Balarajan, Meera (2011) *Exceptional People: How Migration Shaped Our World and Will Define our Future*, Princeton, NJ: Princeton University Press.

Goodin, Robert E. (1992) If people were money..., in Barry, Brian, and Goodin, Robert E. (eds.) *Free Movement: Ethical Issues in the Trans-*

national Migration of People and Money, pp. 6–22, London: Harvester Wheatsheaf.

Goodin, Robert E. (2008) What is so special about our fellow countrymen?, in Pogge, Thomas, and Moellendorf, Darrel (eds.) *Global Justice: Seminal Essays*, pp. 255–284, St. Paul: Paragon House.

Goodwin-Gil, Guy S. (2014) The international law of refugee protection, in Fiddian-Quamiyeh, Elena, Loescher, Gil, Long, Katy, and Sigona, Nando (eds.) *The Oxford Handbook of Refugee and Forced Migration Studies*, pp. 36–47, Oxford: Oxford University Press.

Greene, Joshua (2013) *Moral Tribes: Emotion, Reason, and the Gap Between Us and Them*, London: Penguin.

Griffin, James (2008) *On Human Rights*, Oxford: Oxford University Press.

Haidt, Jonathan (2012) *The Righteous Mind: Why Good People Are Divided by Politics and Religion*, London: Penguin.

Haker, Hille (2002) Identität, in Düwell, Markus, Hübenthal, Christoph, and Werner, Micha H. (eds.) *Handbuch Ethik*, pp. 394–399, Stuttgart: J.B. Metzler.

Hammerstad, Anne (2014) The securitization of forced migration, in Fiddian-Quamiyeh, Elena, Loescher, Gil, Long, Katy, and Sigona, Nando (eds.) *The Oxford Handbook of Refugee and Forced Migration Studies*, pp. 265–275, Oxford: Oxford University Press.

Hodgkinson, David, Burton, Tess, Anderson, Heather, and Young, Lucy (2012) "The hour when the ship comes in": A convention for persons displaced by climate change, in Leckie, Scott, Simperingham, Ezekiel, and Bakker, Jordon (eds.) *Climate Change and Displacement Reader*, pp. 289–326, New York: Earthscan.

Hoesch, Matthias (2016) Allgemeine Hilfspflicht, territoriale Gerrechtigkeit und Wiedergutmachung: Drei Kriterien für eine faire Verteilung von Flüchtlingen – und wann sie irrelevant werden, in Grundmann, Thomas, and Stephan, Achim (eds.) *"Welche und wie viele Flüchtlinge sollen wir aufnehmen?" Philosophische Essays*, pp. 15–29, Stuttgart: Reclam.

Hof, Andries F., de Bruin, Kelly C., Dellink, Rob B., den Elzen, Michel G.J., and van Vuuren, Detlef P. (2009) The effect of different mitigation strategies on international financing of adaptation, *Environmental Science & Policy*, 12 (7), pp. 832–843.

Houghton, John T. (2015) *Global Warming: The Complete Briefing*, 5, Cambridge: Cambridge University Press.

Hsiang, Solomon M., and Burke, Marshall (2013) Climate, conflict, and social stability: What does the evidence say?, *Climate Change*, 123 (1), pp. 39–55.

Huggel, Christian, Wallimann-Helmer, Ivo, Stone, Dáithí, and Cramer, Wolfgang (2016) Reconciling justice and attribution research to advance climate policy, *Nature Climate Change*, 6 (10), pp. 901–908.

Hugo, Graeme (2010) Climate change-induced mobility and the existing migration regime in Asia and the Pacific, in McAdam, Jane (ed.) *Climate Change and Displacement: Multidisciplinary Perspectives*, pp. 9–35, Oxford: Hart Publishing.

International Fund for Agricultural Development (2011) *Weather Index-Based Insurance in Agricultural Development*, available at https://www.ifad.org/documents/10180/2a2cf0b9-3ff9-4875-90ab-3f37c2218a90 (accessed March 1, 2017).

International Organization for Migration (2016) *Global Migration Trends 2015. Factsheet*, available at https://publications.iom.int/system/files/global_migration_trends_2015_factsheet.pdf (accessed March 3, 2017).

— (2017) *Migration and Climate Change*, available at http://www.iom.int/migration-and-climate-change-0 (accessed January 28, 2017).

IPCC (2014a) *Climate Change 2014: Impacts, Adaptation, and Vulnerability. Part A: Global and Sectoral Aspects. Contribution of Working Group II to the Fifth Assessment Report of the Intergovernmental Panel on Climate Change*, available at http://www.ipcc.ch/pdf/assessment-report/ar5/wg2/WGIIAR5-PartA_Final.pdf (accessed March 13, 2017).

— (2014b) *Synthesis Report: Summary for Policy Makers*, available at https://www.ipcc.ch/pdf/assessment-report/ar5/syr/AR5_SYR_FINAL_SPM.pdf (accessed January 26, 2017).

Jones, Sam (2014) 1.7m Syrian refugees face food crisis as UN funds dry up, *The Guardian*, December 1, available at http://www.theguardian.com/world/2014/dec/01/syrian-refugees-food-crisis-un-world-programme (accessed January 23, 2017).

Judis, John B. (2016) *The Populist Explosion: How the Great Recession Transformed American and European Politics*, New York: Columbia.

Kälin, Walter (2010) Conceptualizing climate-induced displacement, in McAdam, Jane (ed.) *Climate Change and Displacement: Multidisciplinary Perspectives*, pp. 81–103, Oxford: Hart Publishing.

Karnein, Anja (2015) Putting fairness in its place: Why there is a duty to take up the slack, *The Journal of Political Philosophy*, 111 (11), pp. 593–607.

Kelman, Ilan (2008) Island evacuation, *Forced Migration Review*, 31, pp. 20–21.

Kniveton, Dominic, Schmidt-Verkerk, Kerstin, Smith, Christopher, and Black, Richard (2008) *Climate Change and Migration: Improving*

Methodologies to Estimate Flows, IOM Migration Research Series, Geneva: International Organization for Migration.

Kolers, Avery (2012) Floating provisos and sinking islands, *Journal of Applied Philosophy*, 29 (4), pp. 333–343.

Kolmannskog, Vikram (2012) Climate changed: People displaced, in Leckie, Scott, Simperingham, Ezekiel, and Bakker, Jordan (eds.) *Climate Change Displacement Reader*, pp. 37–54, New York: Earthscan.

Koser, Khalid (2008) Gaps in IDP protection, *Forced Migration Review*, 31, p. 17.

Kukathas, Chandran (2005) The case for open immigration, in Cohen, Andrew I., and Wellman, Christopher Heath (eds.) *Contemporary Debates in Applied Ethics*, pp. 207–220, Malden: Wiley Blackwell.

— (2016) Are refugees special?, in Fine, Sarah, and Ypi, Lea (eds.) *Migration in Political Theory: The Ethics of Movement and Membership*, pp. 249–268, Oxford: Oxford University Press.

Kymlicka, Will (1989) *Liberalism, Community and Culture*, Oxford: Oxford University Press.

— (1995) *Multicultural Citizenship*, Oxford: Oxford University Press.

— (2001) Territorial boundaries: A liberal egalitarian perspective, in Miller, David, and Hashmi, Sohail (eds.) *Boundaries and Justice: Diverse Ethical Perspectives*, pp. 249–275, Princeton, NJ: Princeton University Press.

Leggewie, Claus, and Welzer, Harald (2009) *Das Ende der Welt, wie wir sie kannten. Klima, Zukunft und die Chancen der Demokratie*, Frankfurt am Main: Fischer.

Leighton, Michelle (2011) Drought, desertification and migration: Past experiences, predicted impacts and human rights issues, in Piguet, Etienne, Pécoud, Antoine, and De Guchteneire, Paul (eds.) *Migration and Climate Change*, pp. 331–358, Cambridge: Cambridge University Press.

Lichtenberg, Judith (2014) *Distant Strangers: Ethics, Psychology, and Global Poverty*, New York: Cambridge University Press.

Linnerooth-Bayer, Joanne, *et al.* (2009) Insurance, developing countries and climate change, *The Geneva Papers*, 34, pp. 381–400.

Locke, John (1988) Two treatises of government, in Lalsett, Peter (ed.) *Cambridge Studies in the History of Political Thought*, Cambridge: Cambridge University Press.

Lyster, Rosemary (2015) Protecting the human rights of climate displaced persons: The promise and limits of the United Nations Framework Convention on Climate Change, in Grear, Anna, amd Kotzé, Louis J. (eds.) *Resarch Handbook on Human Rights and the Environment*, pp. 423–448, Cheltenham: Edward Elgar Publishing.

MacIntyre, Alasdaire (1984) *After Virtue: A Study in Moral Theory*, Notre Dame: Notre Dame University Press.
Maguire, Rowena (2017) Governance of climate displacement within the UN climate regime, in Cadman, Tim, Maguire, Rowena, and Sampford, Charles (eds.) *Governing the Climate Change Regime: Institutional Integrity and Integrity Systems*, pp. 118–135, New York: Routledge.
McAdam, Jane (2012) *Climate Change, Forced Migration, and International Law*, Oxford: Oxford University Press.
McAnney, Sheila C. (2012) Sinking islands? Formulating a realistic solution to climate change displacement, *New York University Law Review*, 87 (4), pp. 1172–1209.
McDowell, Moore, Rodney, Thomas, Frank, Robert, and Bernanke, Ben (2006) *Principles of Economics: European Edition*, Maidenhead: McGraw-Hill Education.
McLeman, Robert A. (2013) *Climate and Human Migration: Past Experiences, Future Challenges*, Cambridge: Cambridge University Press.
McMichael, Anthony, McMichael, Celia E., Berry, Helen L., and Bowen, Kathryn (2010) Climate-related displacement: Health risks and responses, in McAdam, Jane (ed.) *Climate Change and Displacement: Multidisciplinary Perspectives*, pp. 191–219, Oxford: Hart Publishing.
Miller, David (1992) Community and citizenship, in Avineri, Shlomo, and de Shalit, Avner (eds.) *Communitarianism and Individualism*, pp. 85–100, New York: Oxford University Press.
— (1995) *On Nationality*, Oxford: Oxford University Press.
— (2005) Immigration: The case for limits, in Cohen, Andrew I., and Wellmann, Christopher H. (eds.) *Contemporary Debates in Applied Ethics*, pp. 193–206, London: Blackwell.
— (2008) Immigrants, nations, and citizenship, *The Journal of Political Philosophy*, 16 (4), pp. 371–390.
— (2016) *Strangers in Our Midst: The Political Philosophy of Immigration*, Cambridge, MA: Harvard University Press.
Miller, David, and Sundas, Ali (2014) Testing the national identity argument, *European Political Science Review*, 6 (2), pp. 237–259.
Moore, Margaret (2001) *The Ethics of Nationalism*, Oxford: Oxford University Press.
— (2015) *A Political Theory of Territory*, Oxford: Oxford University Press.
National Intelligence Council (2008) *Global Trends 2025: A Transformed World*, Washington, DC: Government Printing Office.
Nawrotzki, Raphael J. (2014) Climate migration and moral responsibility, *Ethics, Policy and Environment*, 17 (1), pp. 69–87.

Neumayer, Eric (2006) The environment: One more reason to keep immigrants out?, *Ecological Economics*, 59 (2), pp. 204–207.

Newton, Kenneth (2001) Trust, social capital, civil society, and democracy, *International Political Science Review*, 22 (2), pp. 201–214.

Nickel, James W. (1983) Human rights and the rights of aliens, in Brown, Peter G., Shue, Henry (eds.) *The Border that Joins: Mexican Migrants and U.S. Responsibility*, pp. 31–45, Totowa: Rowman and Littlefield.

Nine, Cara (2010) Ecological refugees, states borders, and the Lockean proviso, *Journal of Applied Philosophy*, 27 (4), pp. 359–375.

Nordhaus, William (2013) *The Climate Casino: Risk, Uncertainty, and the Economics for a Warming World*, New Haven, CT: Yale University Press.

North, Douglas (1991) Institutions, *Journal of Economic Perspective*, 5 (1), pp. 97–112.

Oberman, Kieran (2013) Can brain drain justify immigration restrictions?, *Ethics*, 123 (3), pp. 427–455.

Oberman, Kieran (2016) Immigration as a human right, in Fine, Sarah, and Ypi, Lea (eds.) *Migration in Political Theory: The Ethics of Movement and Membership*, pp. 32–56, Oxford: Oxford University Press.

OECD (2013) *Income Inequality, Iceland,* available at https://data.oecd.org/inequality/income-inequality.htm#indicator-chart (accessed January 22, 2017).

Oliver-Smith, Anthony (2011) Sea level rise, local vulnerability and involuntary migration, in *Migration and Climate Change*, pp. 160–187, Cambridge: Cambridge University Press.

O'Neill, Onora (2000) *Bounds of Justice*, Cambridge: Cambridge University Press.

Ostrom, Elinor (1990) *Governing the Commons: The Evolution of Institutions for Collective Action*, New York: Cambridge University Press.

Ott, Konrad (2016) Der "slippery slope" im Schatten der Shoa und die Aporien der bürgerlichen Gesellschaft angesichts der Zuwanderung, *Berliner Theologische Zeitschrift*, 2, pp. 47–77.

– (2016) *Zuwanderung und Moral*, Stuttgart: Reclam.

Ottaviano, Gianmarco, and Peri, Giovani (2012) Rethinking the effect of immigration on wages, *Journal of the European Economic Association*, 10 (1), pp. 152–197.

Owen, David (2016) In loco civitatis: On the normative basis of the institution of refugeehood and responsibilities for refugees, in Fine, Sarah, and Ypi, Lea (eds.) *Migration in Political Theory: The Ethics of Movement and Membership*, pp. 269–289, Oxford: Oxford University Press.

Pellegrino, Gianfranco (2014) Climate refugees: A case for protection, in Paola, Marcello, and Pellegrino, Gianfranco (eds.) *Canned Heat: Ethics and Politics of Global Climate Change*, pp. 193-209, London: Routledge.

Penz, Peter (2010) International ethical responsibilities to "climate change refugees", in McAdam, Jane (ed.) *Climate Change and Displacement: Multidisciplinary Perspectives*, pp. 151-173, Oxford: Hart Publishing.

Perry, Stephen R. (1992) The moral foundations of tort law, *Iowa Law Review*, 77, pp. 449-514.

Pevnick, Ryan (2007) Social trust and the ethics of immigration policy, *The Journal of Political Philosophy*, 17 (2), pp. 146-167.

— (2011) *Immigration and the Constraints of Justice: Between Open Borders and Absolute Sovereignty*, Cambridge: Cambridge University Press.

Piguet, Étienne, Pécoud, Antoice, and de Guchteneire, Paul (2011) Migration and climate change: An overview, *Refugee Survey Quarterly*, 30 (3), pp. 1-23.

Pogge, Thomas (2002) Migration und Armut, in *Was schulden wir Flüchtlingen und Migranten? Grundlagen einer gerechten Zuwanderungspolitik*, pp. 110-126, Wiesbaden: Westdeutscher Verlag.

Posner, Eric A., and Weisbach, David (2010) *Climate Change Justice*, Princeton, NJ: Princeton University Press.

Putnam, Robert D. (2000) *Bowling Alone: The Collapse and Revival of American Community*, New York: Simon & Schuster.

— (2007) E pluribus unum: Diversity and community in the twenty-first century, the 2006 Johan Skytte Lecture, *Scandinavian Political Studies*, 80 (2), pp. 137-174.

Rawls, John (1971) *A Theory of Justice*, Harvard, MA: Harvard University Press.

— (1999) *The Law of Peoples*, Cambridge, MA: Harvard University Press.

Rebetez, Martine (2011) The main climate change forecasts that might cause human displacement, in Piguet, Étienne, Pécoud, Antoince, and de Guchteneire, Paul (eds.) *Migration and Climate Change*, pp. 37-48, Cambridge: Cambridge University Press.

Renaud, Fabrice, Bogardi, Janos, Dun, Olivia, and Warner, Koko (2007) Control, adapt or flee: How to face environmental migration?, *Publication Series of the United Nations University, InterSecTions No 5*, pp. 1-444.

Risse, Mathias (2009) The right to relocation: Disappearing island nations and common ownership of the earth, *Ethics and International Affairs*, 23 (3), pp. 281-300.

Sandel, Michael (1992) The procedural republic and the unencumbered self, in Avineri, Shlomo, and de Shalit, Avner (eds.) *Communitarianism and Individualism*, pp. 12-28, Oxford: Oxford University Press.

Schleussner, Carl-Friedrich, Donges, Jonathan F., Donner, Reik V., and Schellnhuber, Hans Joachim (2016) Armed-conflict risks enhanced by climate-related disasters in ethnically fractionalized countries, *Proceedings of the National Academy of Sciences*, 113 (33), pp. 9216-9221.

Schlothfeld, Stephan (2002) Ökonomische Migration und globale Verteilungsgerechtigkeit, in Märker, Alfredo, and Schlothfeld, Stephan (eds.) *Was schulden wir Flüchtlingen und Migranten? Grundlagen einer gerechten Zuwanderungspolitik*, pp. 93-109, Wiesbaden: Westdeutscher Verlag.

Shacknove, Andrew (1985) Who is a refugee?, *Ethics*, 95 (2), pp. 274-284.

— (1988) American duties to refugees: Their scope and limits, in Gibney, Mark (ed.) *Open Borders? Closed Societies? The Ethical and Political Issues*, pp. 131-149, New York: Greenwood Press.

Shue, Henry (1983) Playing hardball with human rights, *Report From the Center for Philosophy and Public Policy*, 3 (4), pp. 9-11.

— (1988) Mediating duties, *Ethics*, 98 (4), pp. 687-704.

— (1996) *Basic Rights: Subsistence, Affluence, and U.S. Foreign Policy*, 2nd ed., Princeton, NJ: Princeton University Press.

— (2004) Thickening convergence: Human rights and cultural diversity, in Chatterjee, Deen K. (ed.) *The Ethics of Assistance: Morality and the Distant Needy*, pp. 217-241, Cambridge: Cambridge University Press.

— (2015) Historical responsibility, harm prohibition, and preservation requirement: Core practical convergence on climate change, *Moral Philosophy and Politics*, 2, pp. 7-31.

Singer, Peter, and Singer, Renata (1988) The ethics of refugee policy, in Gibney, Mark (ed.) *Open Borders? Closed Societies? The Ethical and Political Issues*, pp. 112-130, New York: Greenwoord Press.

Sitter-Liver, Beat (2003) *Gerechte Organallokation. Zur Verteilung knapper Güter in der Transplantationsmedizin*, Fribourg: Academic Press.

Smith, Dan, and Vivekananda, Janani (2012) A climate of conflict: The links between climate change, peace and war, in Leckie, Scott, Simperingham, Ezekiel, and Bakker, Jordan (eds.) *Climate Change and Displacement Reader*, pp. 97-124, New York: Earthscan.

Soroka, Stuart, Banting, Keith, and Johnston, Richard (2006) Immigration and redistribution in a global era, in Bardhan, Pranab, Bowles,

Samuel, and Wallerstein, Michael (eds.) *Globalization and Egalitarian Redistribution*, pp. 261–288, Princeton, NJ: Princeton University Press.

Souter, James (2014) Towards a theory of asylum as reparation for past injustice, *Political Studies*, 62 (2), pp. 326–342.

Stavropoulou, Maria (2008) Drowned in definitions?, *Forced Migration Review*, 31, pp. 11–12.

Steigleder, Klaus (1992) *Die Begründung des moralischen Sollens. Studien zur Möglichkeit einer normativen Ethik*, Tübingen: Attempto.

— (2002) *Kants Moralphilosophie. Die Selbstbezüglichkeit reiner praktischer Vernunft*, Stuttgart: Metzler.

— (2012) *Risk and Rights: Towards a Rights-based Risk Ethics (Working Paper)*, available at http://www.ruhr-uni-bochum.de/philosophy/mam/ethik/content/steigleder_risk_and_rights.pdf (accessed January 3, 2017).

— (2016) Climate risks, climate economics, and the foundations of rights-based risk ethics, *Journal of Human Rights*, 15 (2), pp. 251–271.

Stern, Nicholas (2014) Ethics, equity and the economics of climate change, paper 2: Economics and politics, *Economics and Philosophy*, pp. 445–501.

The Economist (2016) Arab youth: Look forward in anger, August 6.

Thomson, Judith (1986) Some ruminations on rights, in Parent, William (ed.) *Rights, Restitution, and Risks: Essays in Moral Theory*, pp. 49–65, Cambridge, MA: Harvard University Press.

Truong, Thanh, and Gasper, Des (2011) Transnational migration, development and human security, in Truong, Thanh, and Gasper, Des (eds.) *Transnational Migration and Human Security: The Migration-Development-Security Nexus*, pp. 3–22, Berlin: Springer.

Turton, David (2003) Refugees, forced resettlers and "other forced migrants": Towards a unitary study of forced migration, *(UNHCR) New Issues in Refugee Research*, No. 94.

UNFCCC (1992) *Convention Establishing the United Nations Framework Conference on Climate Change*, available at https://unfccc.int/resource/docs/convkp/conveng.pdf (accessed April 3, 2017).

— (2014a) *Climate Finance*, available at http://unfccc.int/cooperation_and_support/financial_mechanism/items/2807.php (accessed March 3, 2017).

— (2014b) *Warsaw International Mechanism for Loss and Damage associated with Climate Change Impacts*, available at http://unfccc.int/adaptation/workstreams/loss_and_damage/items/8134.php (accessed February 13, 2017).

UNHCR (1951, 1967) *Convention and Protocol Relating to the Status of Refugees*, available at http://www.unhcr.org/3b66c2aa10.html (accessed March 15, 2017).
— (2006) *The State of the World's Refugees: Human Displacement in the New Millenium*, Oxford: Oxford University Press.
— (2008) *Statistical Yearbook 2008*, Geneva: UNHCR.
— (2015) *State Parties to the 1951 Convention Relating to the Status of Refugees and Its 1967 Protocol*, available at http://www.unhcr.org/protect/PROTECTION/3b73b0d63.pdf (accessed February 23, 2017).
— (2016) *Global Trends: Forced Displacement in 2015*, available at http://www.unhcr.org/statistics/unhcrstats/576408cd7/unhcr-global-trends-2015.html (accessed March 23, 2017).
Waldon, Jeremy (1995) Minority cultures and the cosmopolitan alternative, in Kymlicka, Will (ed.) *The Rights of Minority Cultures*, pp. 93–119, Oxford: Oxford University Press.
Walzer, Michael (1983) *Spheres of Justice: A Defense of Pluralism and Equality*, New York: Basic Books.
Warner, Koko, Afifi, Tamer, Kälin, Walter, Leckie, Scott, Ferris, Beth, Martin, Susan F., and Wrathall, David J. (2013) *Changing Climate, Moving People: Framing Migration, Displacement and Planned Relocation. Policy Brief No. 8*, Bonn: United Nations University Institute for Environment and Human Security.
Weber, Max (2004) Politics as a vocation, in Owen, David, and Strong, Tracy B. (eds.) *Max Weber: The Vocation Lectures*, pp. 32–93, Indianapolis, IN: Hackett Publishing Company.
Weitner, Thomas (2013) *Menschenrechte, besondere Pflichten und globale Gerechtigkeit. Eine Untersuchung zur moralischen Rechtfertigung von Parteilichkeit gegenüber Mitbürgern*, Münster: Mentis.
Wellman, Christopher H. (2008) Immigration and freedom of association, *Ethics*, 119 (1), pp. 109–141.
Wellman, Christopher H. (2016) Freedom of movement and the rights to enter and exit, in Fine, Sarah, and Ypi, Lea (eds.) *Migration in Political Theory: The Ethics of Movement and Membership*, pp. 80–101, Oxford: Oxford University Press.
Wellman, Christopher H., and Cole, Philip (2011) *Debating the Ethics of Migration: Is There a Right to Exclude?*, New York: Oxford University Press.
Welzer, Harald (2010) *Klimakriege. Wofür im 21. Jahrhundert getötet wird*, Frankfurt am Main: Fischer.
White, Gregory (2011) *Climate Change and Migration: Security and Borders in a Warming World*, Oxford: Oxford University Press.

Williams, Angela (2008) Turning the tide: Recognizing climate change refugees in international law, *Law & Policy*, 30 (4), pp. 502–529.

Williams, Michael C. (2003) Words, images, enemies: Securitization and international politics, *International Studies Quarterly*, 47 (4), pp. 511–531.

World Bank (2014) *Migration and Remittances: Recent Developments and Outlook", Migration and Development Brief, No. 22*, available at http://siteresources.worldbank.org/INTPROSPECTS/Resources/334934-1288990760745/MigrationandDevelopmentBrief22.pdf (accessed January 17, 2017).

— (2016) *Migration and Remittances Factbook 2016, Third Edition*, available at https://openknowledge.worldbank.org/handle/10986/23743 (accessed March 22, 2017).

Wyman, Katrina M. (2012) Are we morally obligated to assist climate change migrants?, *Law & Ethics of Human Rights*, 7 (2), pp. 185–212.

— (2013) Responses to climate migration, *Harvard Environmental Law Review*, 37 (1), pp. 167–216.

Young, Iris Marion (2005) Responsibility and global justice: A social connection model, *Anales de la Cátedra Francisco Suarez*, 39, pp. 709–726.

— (2011) *Responsibility for Justice*, Oxford: Oxford University Press.

Zakaria, Fareed (2016) Populism on the march: Why the West is in trouble, *Foreign Affairs*, 6, pp. 9–15.

Zetter, Roger (2010) Protecting people displaced by climate change: Some conceptual challenges, in McAdam, Jane (ed.) *Climate Change and Displacement: Multidisciplinary Perspectives*, pp. 132–150, Oxford: Hart Publishing.

Index

Absorptive capacity (of societies) 111, 189
Adaptive capacity (of communities or individuals to climate change) 125, 129, 142, 144, 175
Ability to Pay Principle 178
Adaptation Fund (AF) 172
Adelman, Sam 193
Adler, Matthew D. 47
Alesina, Alberto 21
Anderson, Benedict 56, 80
Angeli, Oliviero 92
Arendt, Hannah 44
Associative Ownership View 53, 84–87
Asylum seekers 2, 5, 12, 14–15, 98–90, 130

Baatz, Christian 166, 169, 171, 174
Bader, Veit 17, 91
Banting, Keith 22, 78
Barbieri, Alisson F. 127
Barnett, Jon 123, 127–128, 143–144, 175
Basic rights 5, 27–51, 53, 61, 91–116, 140–146, 165, 178, 187–189
Beitz, Charles 31, 45, 56
Beneficiary Pays Principle 178
Bettini, Giovanni 131
Betts, Alexander 12, 110, 128
Biermann, Frank 177
Black, Richard 150–151
Blake, Michael 18, 93, 94
Boas, Ingrid 131, 177
Bodvarsson, Örn B. 18
Boehm, Peter 128
Borjas, George 18, 19
Brain drain 18, 57
Brezger, Jan 10

Brown, Oli 123, 127
Buchanan, Allen 38, 42
Buhaug, Halvard 129

Cafaro, Philip 23
Campbell, John 128, 183
Caney, Simon 29
Carballo, Manuel 127
Carens, Joseph 14, 52–66, 93, 104, 107–114, 195
Cassee, Andreas 53, 59, 67
Castles, Stephen 14, 149
Categorization of climate migrants 4, 135–145
Causal responsibility 110, 111, 164, 173, 192
Chaturvedy, Sanjay 125, 132
Climate displaced persons (CDPs) 137, 140, 142, 144, 168, 183, 190
Climate refugee treaty 177
Cole, Philip 52, 57
Coleman, James S. 20
Coleman, Jules L. 46, 47, 161–164
Collective resettlement 179–183
Collier, Paul 19–22, 56, 66–69, 80–84, 110, 116, 117, 188
Communitarianism 29, 69–75, 187
Compensation 45–48, 89–90, 148, 153, 155–184
Context of choice 72–80, 112, 180–188
Convention Relating to the Status of Refugees (The 1951 Refugee Convention) 5, 12–14, 52, 94, 130, 149, 152, 177, 192
Cooperation (Social cooperation) 20–22, 45, 114–116, 121
Cooperation games 22, 84, 116, 117, 189

Corrective justice 5, 7, 33, 44–51, 89–90, 106–109, 147, 153–170, 178, 184–191
Cosmopolitanism 7, 9, 20, 53–68, 80, 82, 188
Cournil, Christel 2, 177
Cronin, Adian A. 126
Culture (the relevance of) 8, 40, 53, 66, 69–80, 112, 126, 182, 187

de Shalit, Avner 180, 184
Demography 148, 155
Developed countries 15–18, 109, 154
Developing countries 2, 18, 55, 109, 110, 123, 156, 170, 172
Di Fabio, Udo 42
Dietrich, Frank 181–184, 193
Disjuncture between internal and international movement 58, 59, 62–65
Dispositional freedom 29–30, 64, 95–98, 105, 137, 140–143, 180, 190, 193
Distribution of refugees 109–113
Diversity 20–21, 77, 84
Docherty, Bonnie 13, 177
Donnelly, Jack 39, 61
Doyle, Timothy 125, 132
Drivers of climate migration 123–130
Dummett, Michael 53, 105
Dustman, Christian 19
Dyer, Gwynne 196

Eastern Europe 77, 79, 86, 187
Economic effects of migration 17–19
Elliott, Loraine 129
Emanuel, Kerry 124
Epiney, Astrid 177
Equality of rights 5, 11, 27–33, 50, 54, 92, 101, 116, 161
Ethic of responsibility 11
Ethic of ultimate ends 11
European Union (EU) 1, 99
Exclusive identities 69–72, 77, 79

Finkelstein, Claire 47

Fluidity of categories (of climate migrants) 134–145
Forced migration 2, 4, 16, 124, 138–145, 166, 176, 184, 190
Foresight Report 148
Free movement 3–5, 53–68, 188
Fukuyama, Francis 20

Gardiner, Stephen M. 122
Gasper, Des 132
General causation 155–159
General responsibility 163–167, 179
Gewirth, Alan 28–38, 60, 72, 95, 99, 100, 187
Giannini, Tyler 177
Gibney, Matthew J. 13–16, 94, 97, 145
Gilman, Nils 129–131
Glaeser, Edward L. 21
Globalization 17, 19
Goldin, Ian 12, 14, 20, 22
Goodin, Robert E. 17, 81, 82, 91, 167
Goodwin-Gil, Guy S. 12
Greene, Joshua 57, 116
Griffin, James 35
Guatemala 151–152
Guiding Principles on Internal Displacement 177

Haidt, Jonathan 116
Haker, Hille 74
Hammerstad, Anne 130
Harm (and compensation) 45–49
Hobbes, Thomas (Hobbesian state of nature) 20
Hodgkinson, David 156, 177
Hoesch, Matthias 115–117
Hof, Andries F. 169
Houghton, John T. 125–129
Hsiang, Solomon M. 129
Huggel, Christian 155–157
Hugo, Graeme 103, 128, 136–139, 143–144, 148, 171, 175

Iceland 65–68
Idealism 9, 11
Identity, thin, thick 69–80
In situ assistance (of refugees) 98, 104, 169, 192

Incentives (in the migration regime) 12, 40–41, 88, 149, 176
Inequality 21, 54–58, 65
Infringement (of rights) 32, 46, 64, 161
Insurance (system) 150, 166, 170–178, 192–193
Intergovernmental Panel on Climate Change (IPCC) 121–128, 146
Internal movement 59, 61–64, 176
Internally Displaced Persons (IDPs) 2, 130, 176
International Fund for Agricultural Development 174
International Organization for Migration (IOM) 1, 135–136
Issue of causation 109, 145–155, 174, 184, 191
Issue of multi-causality 184, 191, 148–155

Japan 55–57
Jones, Sam 103
Judis, John B. 1

Kälin, Walter 123–124, 129–130, 135, 138, 145–146
Karnein, Anja 114
Kelman, Ilan 128
Kniveton, Dominic 125
Kolers, Avery 181–183, 193
Kolmannskog, Vikram 176
Koser, Khalid 130
Kukathas, Chandran 17, 104
Kymlicka, Will 22, 70–79, 86

Least Developed Country Fund (LDCF) 172
Legitimacy (of states) 38
Leighton, Michelle 126, 128
Liberalism 54, 73–74
Lichtenberg, Judith 48, 108
Limits of immigration (and of our duty to admit refugees) 6, 93, 114–118, 188, 190
Linnerooth-Bayer, Joanne 170
Locke, John 85
Lyster, Rosemary 13, 137, 170, 173, 176–177

MacIntyre, Alasdaire 67
Maguire, Rowena 158, 172
McAdam, Jane 13, 130–132, 142–148, 154, 176–180
McAnney, Sheila C. 169
McDowell, Moore 17
McLeman, Robert A. 136
McMichael, Anthony 127, 175
McMichael, Celia E. 127, 175
Membership (in a particular political community) 12, 68, 70–73, 87
Miller, David 7–10, 21, 29, 38, 42, 52, 57, 60, 69, 76–79, 98, 109, 179–180, 195
Moore, Margaret 40, 77, 99, 183, 194
Moral Division of Labour (MDL) 33–38, 43, 56, 81
Multi-causality (the issue of) 184, 191, 148–155
Multiculturalism 22
Mutual regard 20, 81, 83, 116

National identity 72, 76–81, 86
Nawrotzki, Raphael J. 3, 146
Negative duties 91, 99, 164
Neumayer, Eric 33–35, 91, 99, 164
Newton, Kenneth 20
Nickel, James W. 35
Nine, Cara 193
Non-refoulement 98–104, 110
Nordhaus, William 2
North, Douglas 67

Oberman, Kieran 18, 58
Occurrent freedom 30, 95–96, 140
Oliver-Smith, Anthony 129
O'Neill, Onora 71
Open borders 3, 16–18, 53–69, 188
Organization for Economic Co-operation and Development (OECD) 65
Ostrom, Elinor 20
Ott, Konrad 59, 66
Ottaviano, Gianmarco 19
Owen, David 38, 104, 111, 113
Oxford armchair style 67

Pellegrino, Gianfranco 193

Penz, Peter 3, 173–176
Perception of risk (as a driver of climate migration) 136–142, 171, 192
Pevnick, Ryan 53, 60, 61, 84–88
Piguet, Étienne 1, 125, 128, 129, 148, 150, 177
Pogge, Thomas 18, 91
Political approach 8–12, 32, 42, 96, 83, 185, 187, 194
Population growth (and migration) 23
Populism 56, 185
Positive duties 33, 82, 99
Posner, Eric A. 154
Potential climate migrants 138, 141–145, 149, 156, 158–159, 165–170, 176, 178, 190, 191
Preventive responsibility 163–170, 181, 185, 191, 193
Problem of attribution 147
Protracted refugee situation 105–106
Putnam, Robert D. 20–21

Racism 19, 71, 76, 86, 188
Rawls, John 40–41, 52, 73, 149
Realism 9, 42
Rebetez, Martine 124–128
Refugees 1–3, 8, 12–14, 30, 38, 47, 51, 89, 92, 93–119, 129–132, 136, 141, 146, 149, 152, 154, 173, 176, 177, 184, 189–190, 192, 194, 195, 196
Remittances 18, 144
Renaud, Fabrice 136
Resettlement 103–106, 179–184
Right to asylum 98–104, 189–190
Right to exclude 52–92, 114–115, 141, 169, 187–188
Right to self-determination 34–42, 64, 85, 88, 115
Right to stay 142–143, 169, 175–176, 193
Risk ethics 11–12, 22, 35, 48, 49–51, 50, 65, 83–84, 92, 114–118, 166–170, 184, 189, 193, 195
Risse, Mathias 181, 193
Rooms of justice 37

Sandel, Michael 69
Schleussner, Carl-Friedrich 129
Schlothfeld, Stephan 91
Sea-level rise (SLR) 121–124, 128–129, 151, 181–183, 187, 193
Securitization (of climate migrants) 6, 130–133, 186, 196
Self-responsibility (of states) 39, 41, 51, 56, 132
Shacknove, Andrew 14, 94
Shue, Henry 10, 11, 28–37, 48, 60, 64, 81, 89, 94–96, 121–122, 178
Singer, Peter 101–102
Singer, Renata 101–102
Sitter-Liver, Beat 37, 66, 74
Slow-onset (climatic) triggers of migration 123–124, 126–130, 134–139, 144–146, 157, 190
Smith, Dan 129
Social effects of migration 5, 16–23
Social trust 8–9, 20–22, 53, 69, 76–77, 80–83, 112, 188
Soroka, Stuart 22
Special duties 5, 33–37, 45–48, 69, 80–88
Specific causation 155–159, 163, 191
Stability (and instability of social cooperation) 22, 56, 64, 84, 117–118, 128, 160, 193
Standard threats 37, 61, 96–98, 193
Stavropoulou, Maria 177
Steigleder, Klaus 28–29, 31–32, 49–51, 115
Sudden-onset (climatic) triggers of migration 123–126, 134–137, 140–146, 157, 174–175

Thomson, Judith 46
Timing factor (its relevance in the context of climate migration) 124, 126, 138–139, 143, 168, 190
Tolerance 8, 9, 194
Trump, Donald 1, 92
Truong, Thanh 17
Turton, David 133

Unemployment 18
United Nations 2, 171, 172, 192
United Nations Development Programme (UNDP) 176

United Nations Framework Convention on Climate Change (UNFCCC) 172–173
United Nations High Commissioner for Refugees (UNHCR) 2, 12–14, 105–106, 109, 173, 176–177

Van den Berg, Hendrik 18
Violation (of rights) 33–34, 38, 44, 46, 48, 61, 64, 101–102
Vivekananda, Janani 129
Voluntary migration 2, 4, 6, 16, 89, 93, 100, 124, 134–136, 138–141, 190
Vulnerability 30, 46, 60, 83, 92, 95–97, 125, 128, 142–144, 157, 161, 163–167, 170–173, 178, 184, 192

Waldron, Jeremy 74
Walzer, Michael 69–70, 75, 101–102, 107, 184
Warner, Koko 175

Webber, Michael 123, 127–128, 143–144, 175
Weber, Max 11
Weisbach, David 154
Weitner, Thomas 31, 34, 81–82
Welfare state 9, 22, 76
Wellman, Christopher H. 52, 57, 60, 68, 98, 195
Welzer, Harald 130
White, Gregory 77, 132
Williams, Michael C. 130, 177
World Bank 18
Wündisch, Joachim 181–184, 193
Wyman, Katrina M. 3, 147, 154

Xenophobia 19, 112

Young, Iris M. 18, 107–108, 160–165

Zakaria, Fareed 1, 8, 19, 21, 56–57, 185
Zetter, Roger 135, 138, 146–147, 154, 183